F UNITS

The diesels that did it

JEFF WILSON

KALMBACH
BOOKS

© 2000 Kalmbach Publishing Co. All rights reserved. This book may not be reproduced in part or in whole without written permission of the publisher, except in the case of brief quotations used in reviews. Published by Kalmbach Publishing Co., 21027 Crossroads Circle, Waukesha, WI 53187.

Printed in the United States of America

99 00 01 02 03 04 05 06 07 08   10 9 8 7 6 5 4 3 2 1

Visit our website at
http://books.kalmbach.com
Secure online ordering available

Publisher's Cataloging in Publication
(Provided by Quality Books, Inc.)

Wilson, Jeff.
    F units : the diesels that did it / Jeff
Wilson. — 1st ed.
    p. cm. — (Golden years of railroading)
    Includes index.
    ISBN: 0-89024-374-3

    1. Diesel locomotives—United States.
I. Title. II. Series.

TJ619.2W55 1999   625.2'66'0973
                                QBI99-927

Book design: Mark Watson • Cover design: Kristi Ludwig

On the Cover: Wabash F7 No. 1160 and Canadian National 2-8-2 No. 3431 await their next assignments at Fort Erie, Ontario, in 1956. Photo by Jim Shaughnessy.

Contents

The FT: The Diesel That Did It 5

F Units in Production:

Inside EMD's La Grange Plant 9

F Units Model by Model 14

F Units Hard at Work 28

F Units Today 124

Dedication 126

Index 127

The FT: The Diesel That Did It

Throughout the 1930s diesel-electric locomotives proved that they could be successful in both high-speed passenger service and yard switching. However, many skeptics remained unconvinced that diesels could do a better job than steam with the meat and potatoes of railroading: heavy-duty road freight service. General Motors subsidiary Electro-Motive Corporation had other ideas and set out to prove the skeptics wrong with the FT.

In the FT, Electro-Motive intended to build a diesel that could be mass-produced—an off-the-shelf model that any railroad could order with only minor variations. It had to be a locomotive suitable for many situations: one equally adept at hauling heavy trains on mountain grades, fast freights on the prairie, and even passenger trains if necessary.

The four-unit streamlined locomotive that rolled out of Electro-Motive's La Grange, Illinois, plant in November 1939 looked familiar, bearing a family resemblance to the many sleek passenger E units that had preceded it. However, the FTs looked more blue-collar and businesslike. The bulldog nose wasn't as slanted and sleek as EMC's contemporary E4 passenger engine, and the FTs looked more compact and powerful.

Each unit was shorter and housed a single 567-series V-16 diesel engine instead of the twin V-12s of the Es. The FT rode on two-axle trucks with all axles powered, whereas the three-axle trucks of the Es had an unpowered center (idler) axle. This put all of the FT's weight atop driving axles. That, combined with a higher gear ratio, gave the FTs greater tractive effort at low speeds than comparable steam locomotives.

Each unit produced 1,350 hp, and Electro-Motive stressed the idea of creating a single locomotive with the building-block approach: an A-B set for 2,700 hp, an A-B-A set for 4,050 hp, or an A-B-B-A set for 5,400 hp.

After leaving the factory the four-unit demonstrator locomotive

◄ Half of the Electro-Motive FT demonstrator, originally numbered 1030 (the other was 1031, the units' serial numbers) but later changed to 103, is shown here in 1939 just before embarking on its nationwide trial. Electro-Motive Corp. photo.

▲ The four-unit demonstrator set pauses at Missoula, Montana, while testing on the Northern Pacific in 1940. Photo by R. V. Nixon.

▲ ▶ The Santa Fe was the first to place an order for the revolutionary FT. Number 100 was the first one delivered. It was completed in December 1940 and entered service in February 1941. Santa Fe photo.

began an 11-month, 83,000-mile working tour of the country's railroads. The FTs rolled freights across hot desert territory, where water for steam locomotives was scarce. They pulled 16,000-ton iron ore trains on Minnesota's iron range and hauled heavy freights up steep Rocky Mountain grades.

By the time the FT had completed its tour, it not only proved that diesels could do the job, but that diesels could do it more cheaply and efficiently than steam locomotives. The Santa Fe, suitably impressed with the way the FTs handled, placed its first FT order in September 1940 and received the first production model, No. 100, in December 1940. The railroad eventually became the largest FT owner with 320 FTs.

The Electro-Motive Corporation became the Electro-Motive Division of General Motors in January 1941. As FT orders came pouring in through 1941, the onset of U. S. participation in World War II had a dramatic impact on locomotive production. Thanks to the success of the FT, EMD was the only builder allowed by the War Production Board to produce road freight diesels during the war. Competitors Alco and Baldwin were limited to making switchers. This gave EMD a huge advantage, essentially a four-year head start to work out bugs and improve the F unit design.

The final FTs, orders for Great Northern and Rock Island, rolled out of La Grange in November 1945. Steam locomotives still outnumbered diesels, but the FT had more than proved the viability of diesels in mainline freight service. The next wave of improved diesels, including EMD's succeeding F units, would spell the end of mainline steam in the 1950s.

Evolution of the F

Continued refinements to the 567 engine, generator, traction motors, and other components led to better reliability, less maintenance, and improved power in succeeding F unit models.

Many FTs were delivered as drawbar-connected A-B sets, but

railroads soon learned the versatility of the building-block idea, so all Fs from the F2 on had couplers at each end.

Multiple-unit (m.u.) connections were originally located on the rear of A units and both ends of B units. The idea was that an A would always lead and trail. This arrangement was rather restrictive, so the availability of m.u. connections on A unit noses became an option. This was marked by a small hatch near the headlight to cover the m.u. socket, as well as m.u. hoses on the pilot. Nose m.u. was later retrofitted to many earlier engines.

Although originally designed as heavy-duty freight locomotives, Fs were available with steam generators for passenger service. Some were freight units intended for emergency passenger use, but many railroads acquired fleets of Fs assigned specifically to passenger service. This was especially true in mountainous areas and for slower passenger trains. The standard gear ratio was 62:15 (65 mph maximum), but several lower ratios were available to 56:21 (102 mph) for passenger service.

The F's popularity as a passenger engine led to the FP7, a stretched version of the F7. The FP7 was four feet longer than an F7, enabling it to carry a higher-capacity steam generator with a larger water tank.

Other F unit options included a passenger-style pilot with retractable coupler as found on EMD E units, dynamic brakes, various types of horns, second (low) headlight, and large side number boards.

As the success of the FT predicted, the newer F3 and F7 pushed steam locomotives out of the American railroad picture. A total of 1,805 F3s and 3,862 F7s were built—far more than the FAs of Alco, the nearest competitor. However, the growing popularity of road switchers such as EMD's GP7 in the early 1950s spelled the end for F units. The Fs were great road locomotives, but their lack of rear visibility made it difficult to do switching on local freights. Also, the streamlined carbody with its complex nose curves was more expensive to produce than the boxy, utilitarian GP.

By the time the F9 and GP9 were introduced in 1954, railroads had turned almost completely to road switchers, and only 516 F9s were built through 1958, compared to more than 3,500 GP9s.

The final F units built were a special order of the New York, New Haven & Hartford. The 60 FL9s were basically FP9s with a third-rail shoe that could be used underground into New York's Grand Central Station. The final F, New Haven FL9 No. 2059, rolled out of La Grange in November 1960, 21 years and 7,690 F units after FT 1030 in 1939.

Although many steam fans of the 1930s and '40s predicted that railfans would have no interest in diesels, the F sparked a new type of interest in trains. The large, smooth sides of F units were like rolling billboards, and designers at EMD and the railroads came up with many colorful and interesting paint schemes that held the interest of both railfans and the public, even now, nearly 40 years since the last F rolled off the line.

F Units in Production: Inside EMD's La Grange Plant

Much of Electro-Motive's groundbreaking early work was done while the company didn't even have its own production facility. The company decided to build its own factory to bring in work that had been farmed out to subcontractors and consolidate production in one facility. Ground was broken at La Grange, Illinois, in March 1935, and the first locomotive completed at the plant emerged on May 20, 1936.

Electro-Motive made an early decision to produce all of its own components, including engines, generators, traction motors, and car-bodies. This meant that GM and EMC were ultimately responsible for everything on the locomotive. It also meant that parts were standardized among models, simplifying parts availability and making assembly-line production more feasible.

Production of the FT stepped up quickly in 1941. However, with the start of World War II, EMD, under direction of the War Production Board, ceased making locomotives for a few months in 1942. Instead, EMD built 567 engines for Navy use. Back in limited locomotive production in late 1942, the plant averaged one FT per day.

As the war ended, EMD increased production to get out the backlog of FT orders, and the plant itself continued to grow. With large orders for F3s and F7s—not to mention E units and switchers—after the war, the plant was eventually expanded to 3.5 million square feet of manufacturing space. By peak F unit production in the early 1950s EMD was turning out almost ten locomotives per day.

These photos were taken at EMD's Plant 1, where the locomotives were assembled and many components were manufactured. Plant 2, on the south side of Chicago, was responsible for heavy steel cutting and fabricating, creating many subassemblies. Another facility, Plant 3 in Cleveland, was

Continued on page 12

◀ Workers use a crane to lower an F unit body onto its trucks. Electro-Motive photo.

◀◀ A long row of F7 noses rolls down the assembly line at Plant 1 next to a lineup of GP7 cab/hood assemblies. The cab assemblies came from Plant 2 on Chicago's south side. Electro-Motive photo, 1951.

◀ A 16-cylinder 567 engine, generator, and electrical cabinet rest on a frame. The carbodies are in place on an F7 A and B unit in the background. Electro-Motive photo, 1952.

▲ Carbody diesels such as F units (this is an F3) had fairly thin frames, relying on truss assemblies in each side for strength. Electro-Motive photo, 1948.

▶ The back lot at EMD looked much like a modeler's workbench, with assembled-but-unpainted F7s sharing space with finished locomotives such as the Denver & Rio Grande Western F7s in the background at left. Electro-Motive photo.

Continued from page 9

responsible for assembling switching and some GP locomotives.

To serve its Canadian customers, EMD opened its Canadian subsidiary, General Motors Diesel Ltd., or GMD, in August 1950. That plant is located in London, Ontario. Together, La Grange and London built all of the nearly 7,700 F units produced.

F Units Model by Model

Ratio 1:87 or HO scale
To convert HO scale drawings to your scale, copy at these percentages: N, 54%; S, 136%; O, 181%.

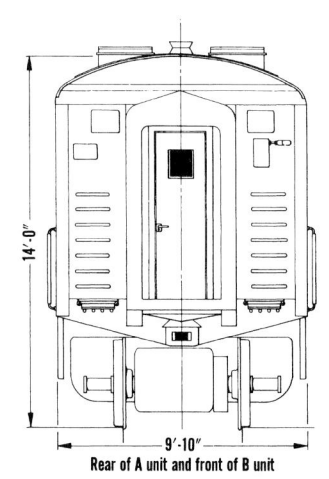

Rear of A unit and front of B unit

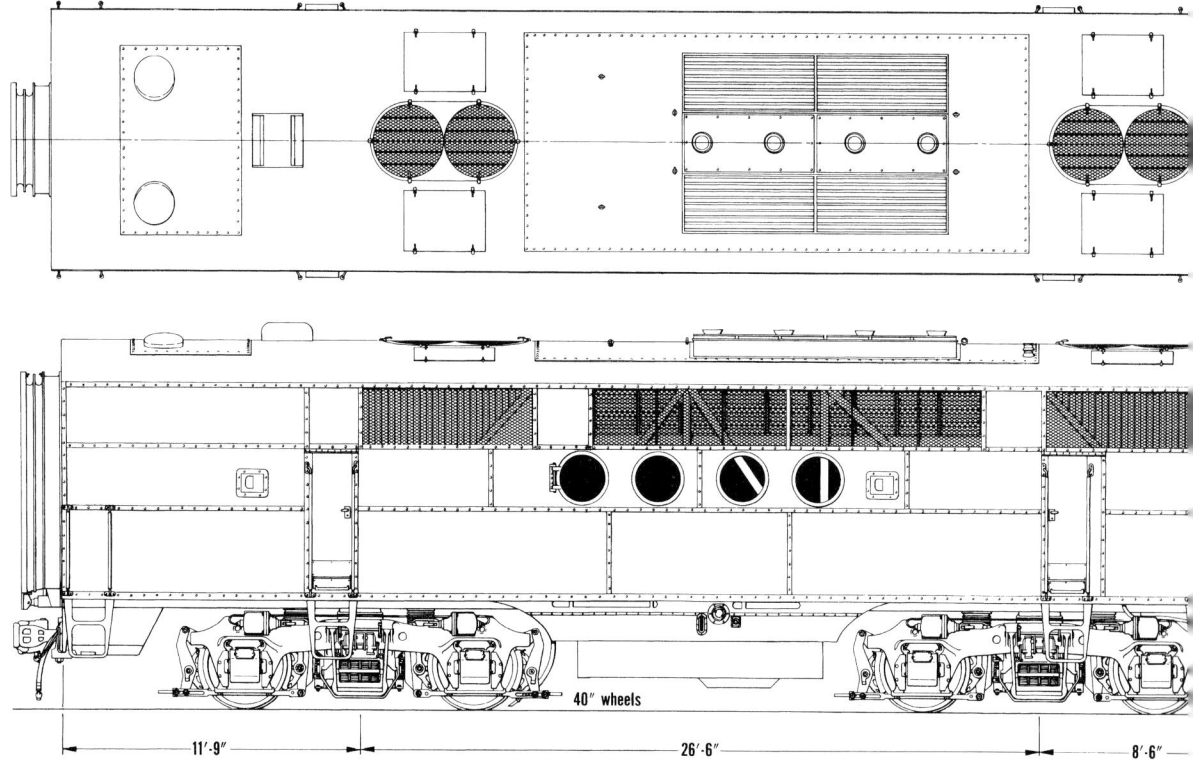

F2

The F2 was a transition model between the FT and F3. The locomotive had the upgraded 567B engine but retained the D8 generator of the FT, resulting in 1,350 hp. The body was the same as the first F3s. A total of 74 A units and 30 B units were built from July to November 1946. These Chicago, Burlington & Quincy units have not yet received their nose heralds.

Chicago, Burlington & Quincy F2 No. 153, built in 1946. Electro-Motive photo.

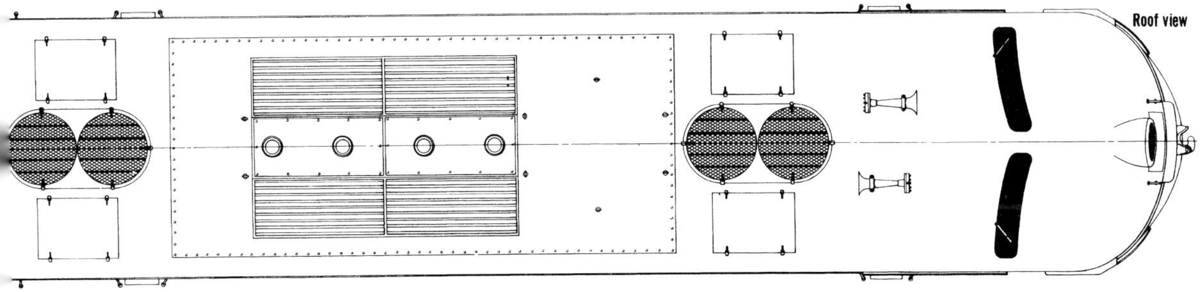

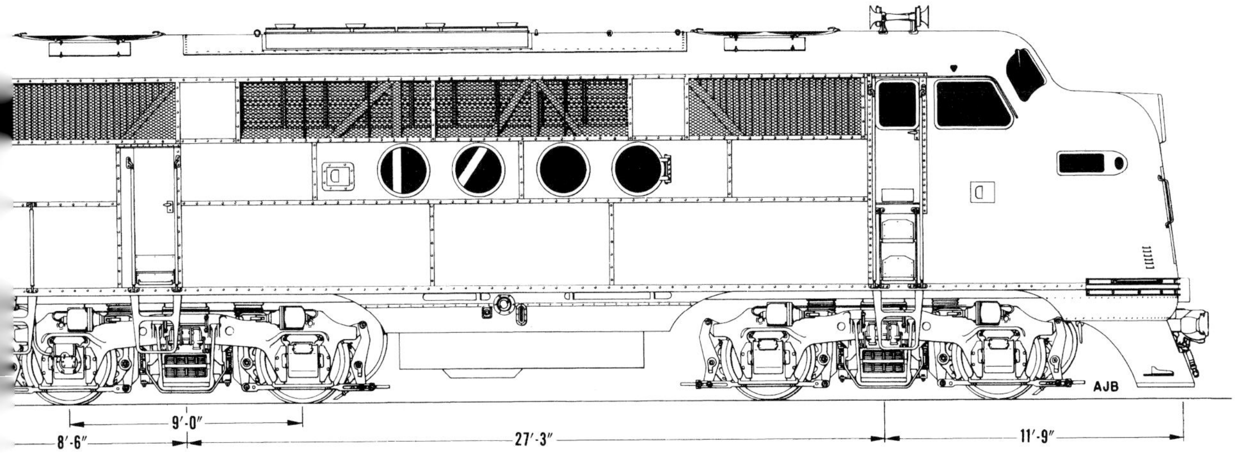

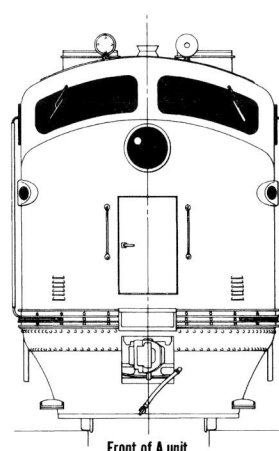

Phase I

All F3s were 1,500-hp locomotives, and a total of 1,111 A units and 696 B units were built from July 1945 through February 1949. The Phase I F3, shown by the drawing below, was characterized by three side portholes and louvered upper side panels covered with screen wire. Dynamic brakes were optional on F3s, and these engines have them, evidenced by the rectangular screen-covered openings on the roof just behind the horns. Phase I F3s were built through May 1947.

Phase II Early

The first major change in the F3's body style was the elimination of the center porthole and the addition of four filtered engine-air-intake openings on the side. The entire area between the two remaining portholes was covered with screen wire. The large corner number boxes became standard on F3s, although the smaller side number boxes remained an option. These F3s were built from June through December 1947.

Phase II Late

The next change came in December 1947 when low-profile rooftop radiator fans replaced the high-shrouded fans of earlier F3s. These were built through July 1948.

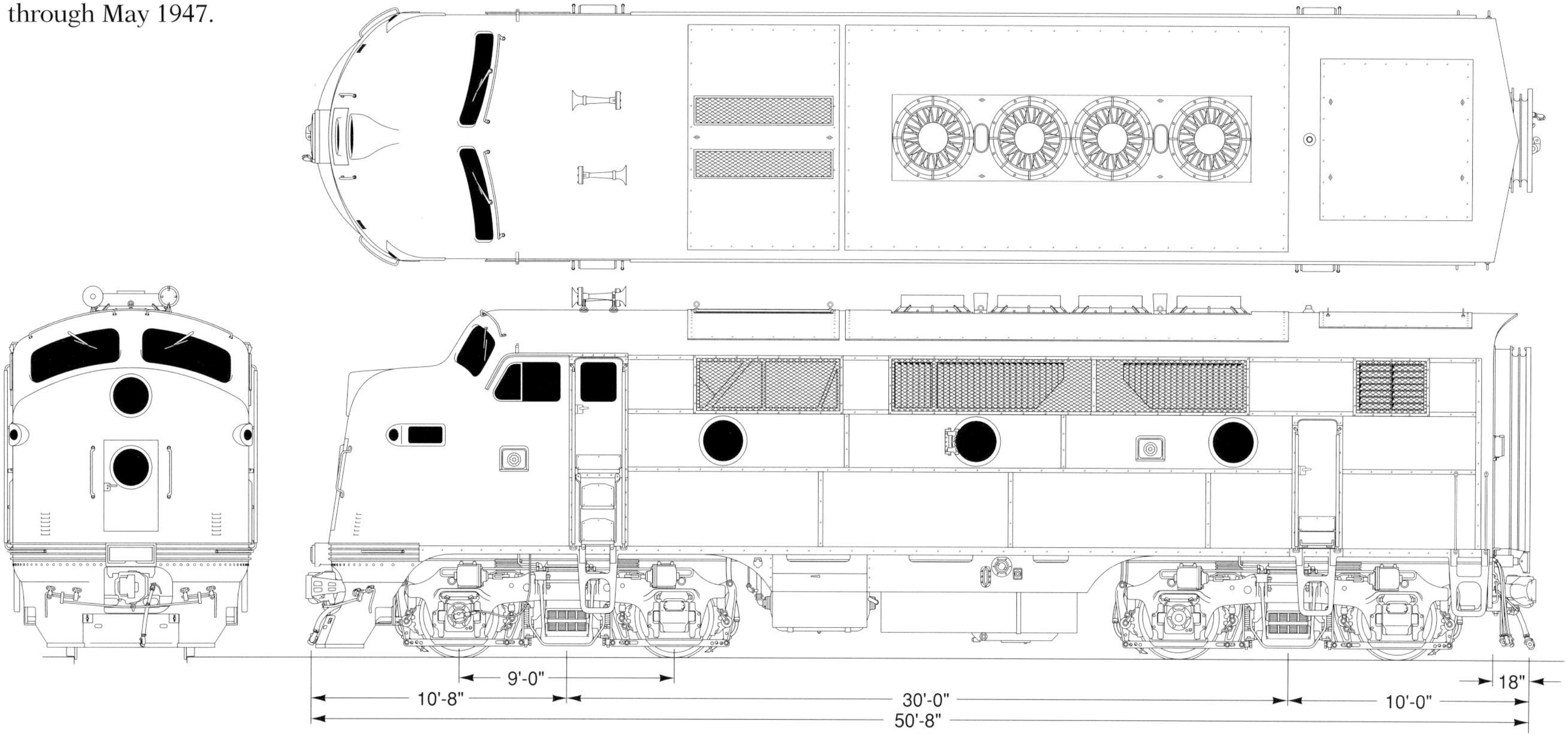

Phase III

Beginning in July 1948 the air-intake filters were louvered, and the screen wire was eliminated between portholes. This body style was produced through September 1948.

Phase IV

In the final body style used on F3s, built from September 1948 through February 1949, stainless steel grills replaced screen wire over the louvers along the upper sides. Phase IV F3s looked almost identical to early F7s, but the Phase IV F3s retained the rectangular roof openings for dynamic brakes.

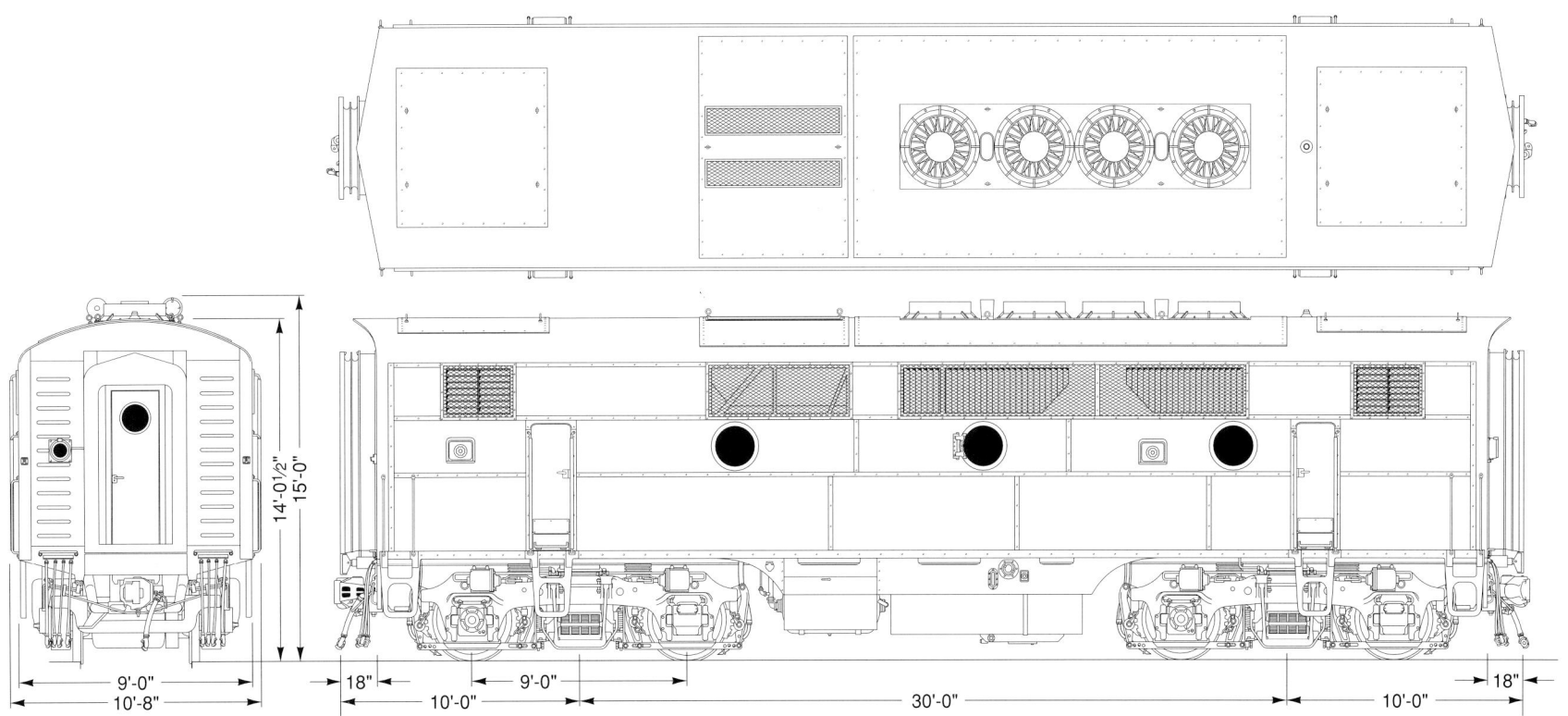

F3 Cutaway

F3 Phase II Early — Electro-Motive photo

F3 Phase II Late — R. H. Payne photo

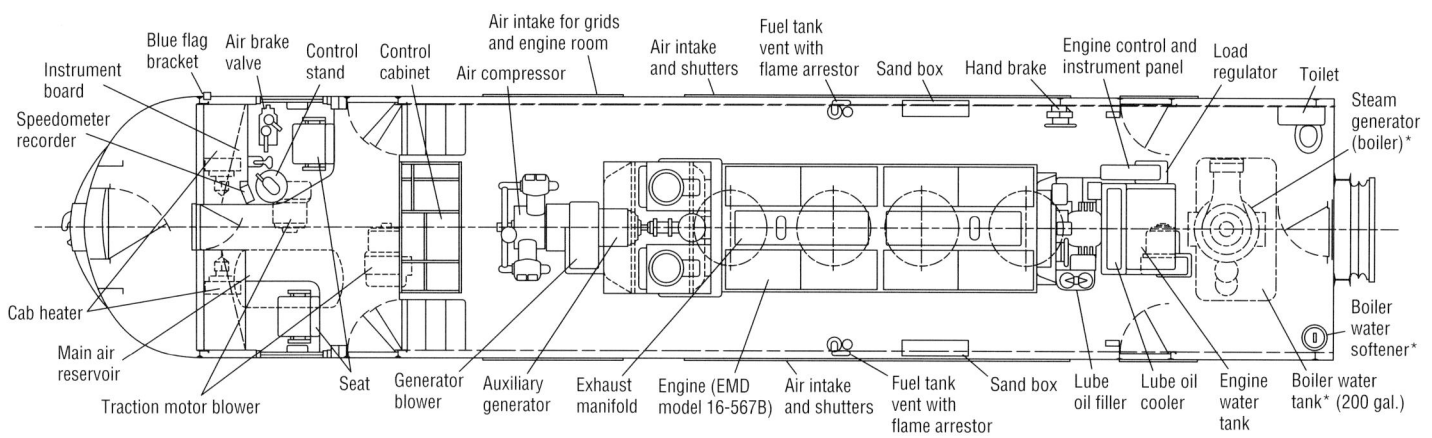

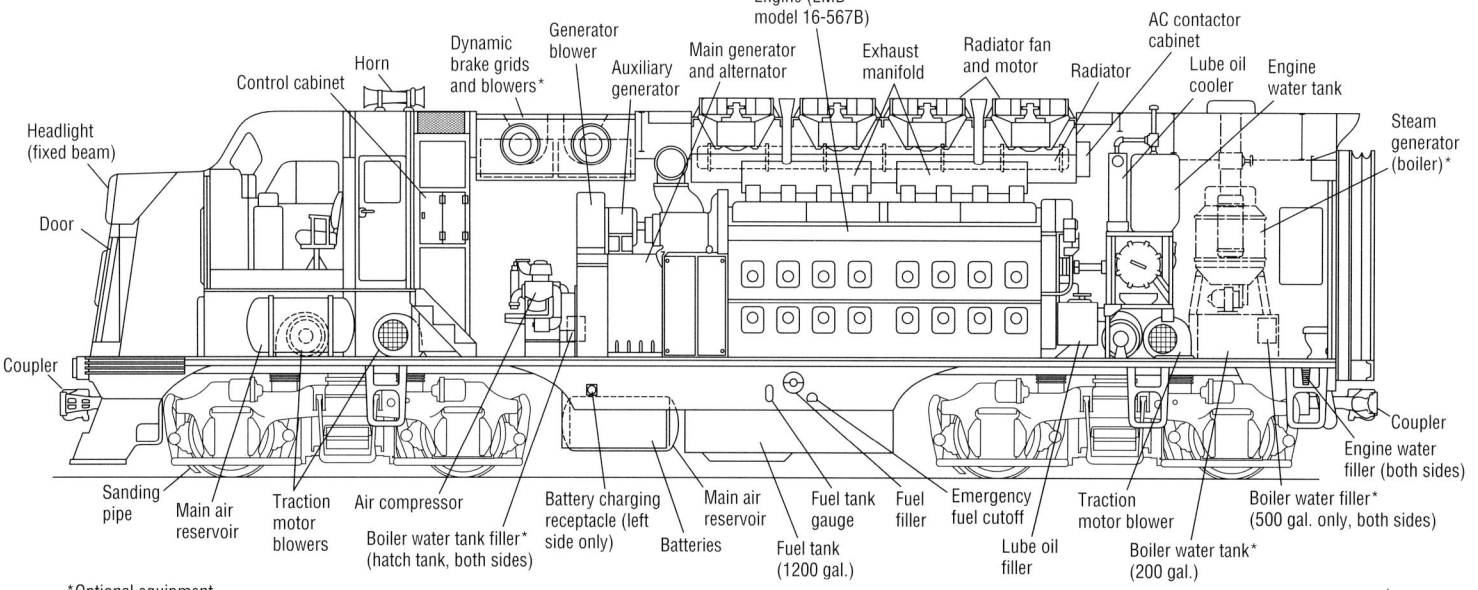

*Optional equipment

F3 Phase III — Missouri Pacific photo

F3 Phase IV — Chicago, Burlington & Quincy photo

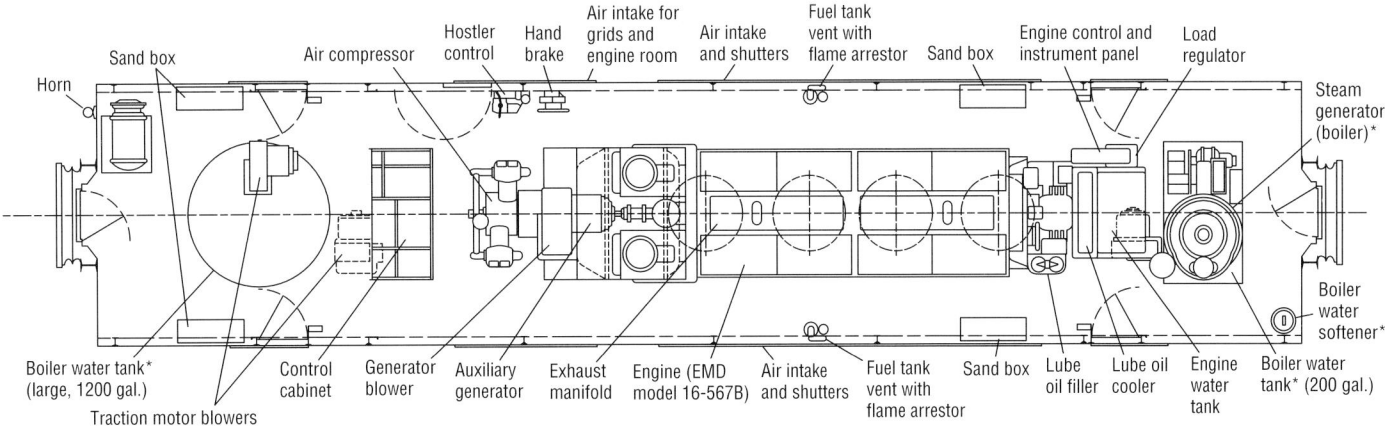

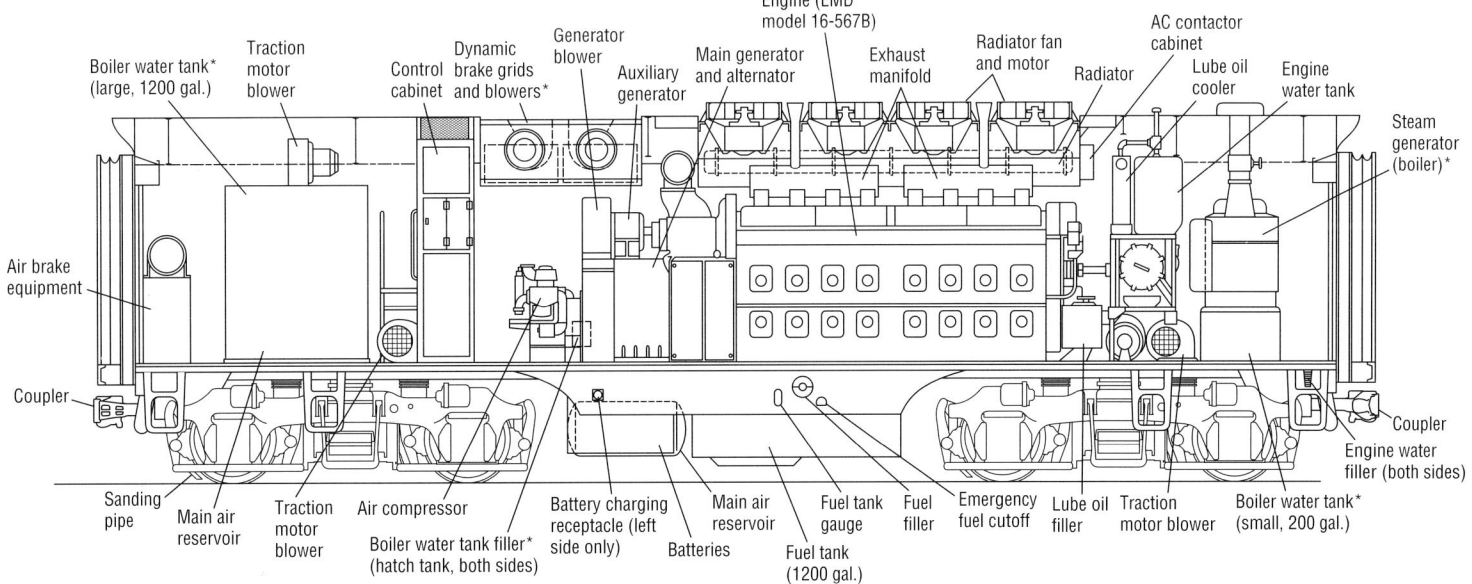

*Optional equipment

Phase I

The drawing below shows the appearance of the F7 from its introduction in November 1948 through August 1952. The only external change from the Phase IV F3 was the addition of a dynamic brake fan (on the roof directly above the forward porthole) in place of the rectangular screened openings shown on page 16. The corners of cab and engine room doors were rounded starting in 1950. Mechanically, F7s were rated at 1,500 hp, the same as F3s, but F7s received the new D27 traction motor, which boosted the F7's tonnage rating about 30 percent. With a total of 2,316 A units and 1,483 B units built, the F7 was the most popular F unit.

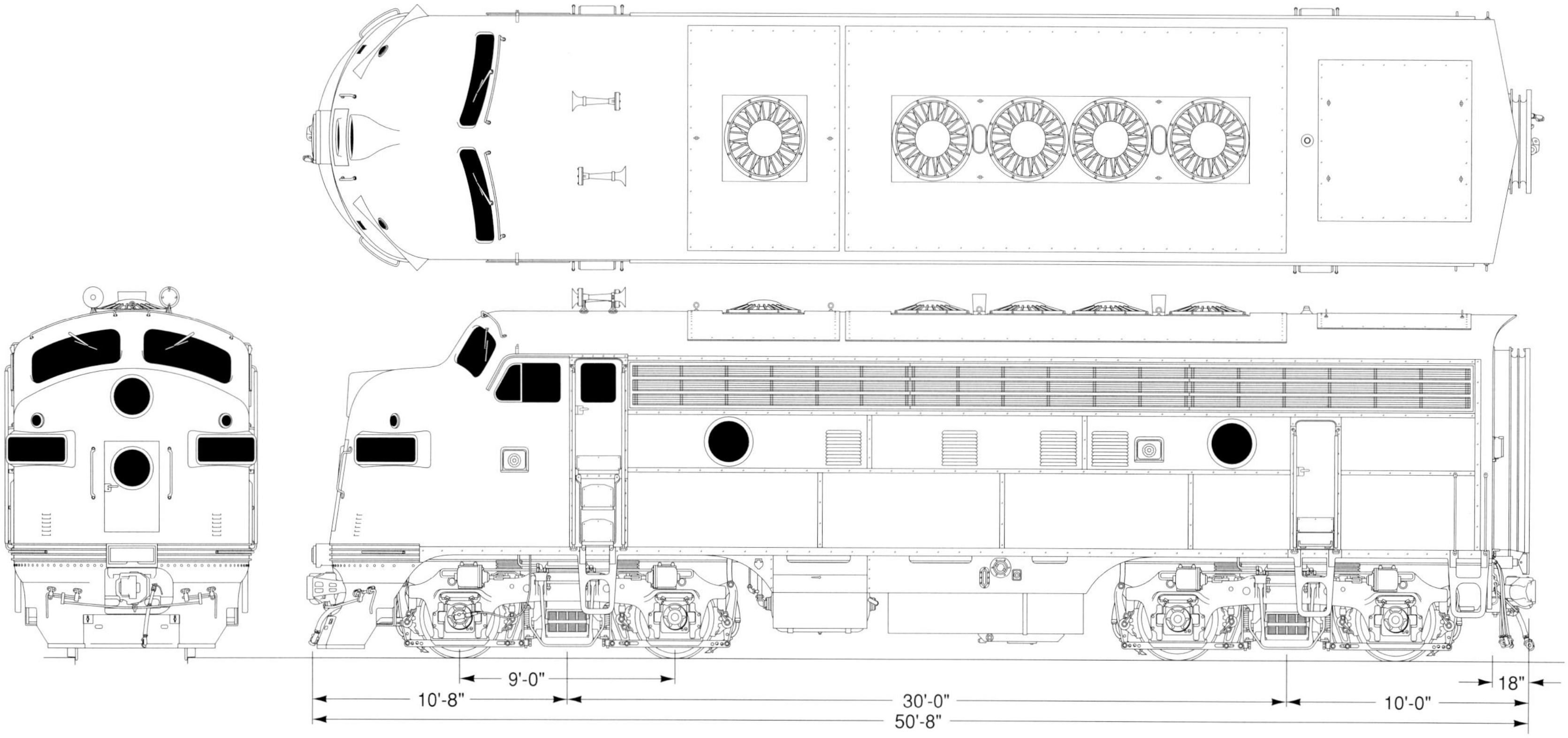

Phase II

Several features changed during F7 production. The first was the introduction of a 48-inch-diameter dynamic brake fan, replacing the former 36-inch fan, in August 1952. Next was a stainless steel Farr-Air grill with vertical slits, starting in November 1951. The old style of louvers gave way to twin pairs of vertical-slit louvers around February 1952. These features remained until the end of F7 production in December 1953.

Chesapeake & Ohio F7 Phase II No. 7076. Photo by Jack Emerick.

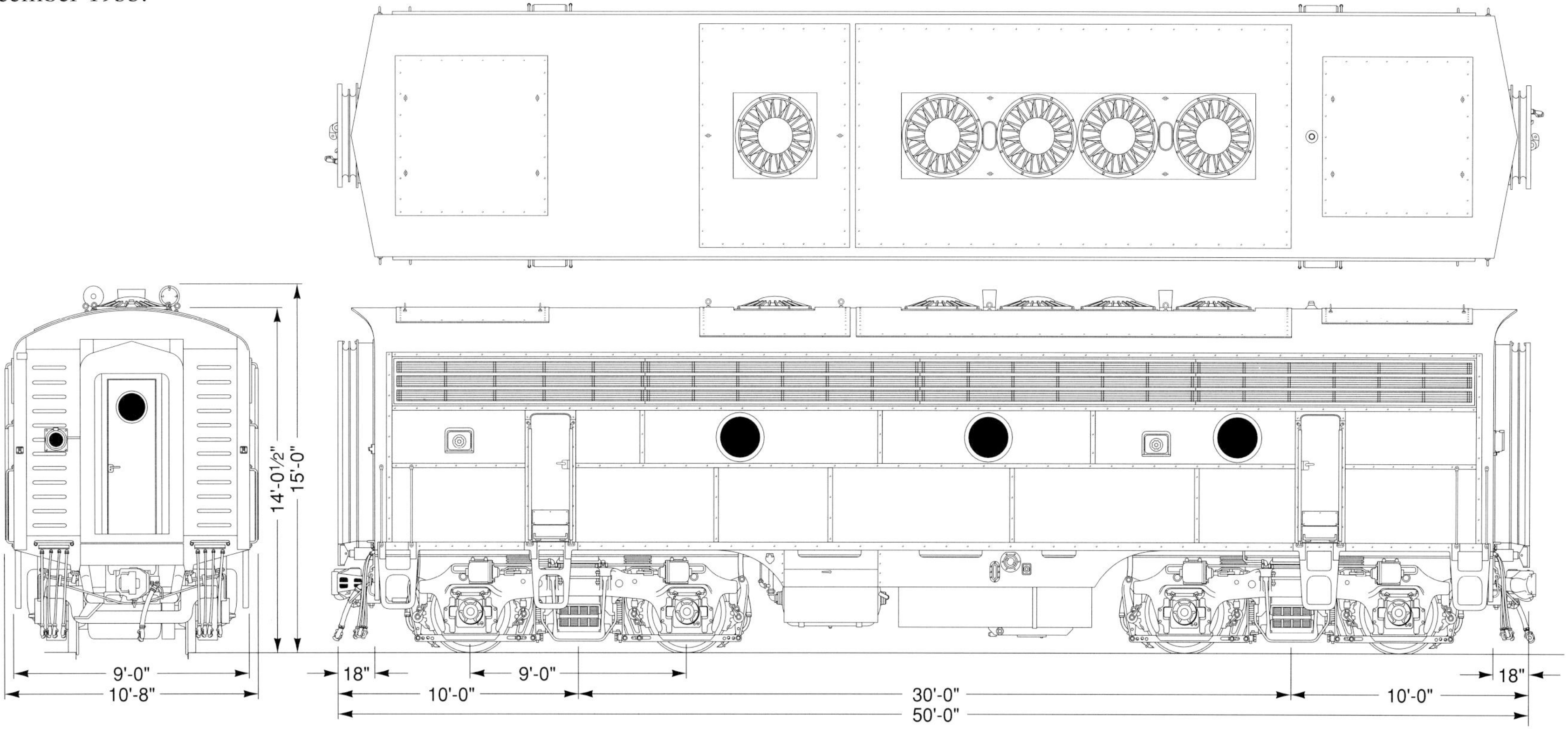

F9

The F9 looked a lot like a late F7, but the forward porthole was moved back and another set of air intake louvers was added just ahead of the porthole. The large (48-inch) dynamic brake fan is apparent in the drawing. The F9 was more powerful than its predecessors, putting out 1,750 hp with its improved 567C engine and D37 traction motors. However, by the time the F9 was introduced in January 1954, railroads had turned toward more versatile hood-type units such as the GP9. Only 87 F9As and 154 F9Bs were built.

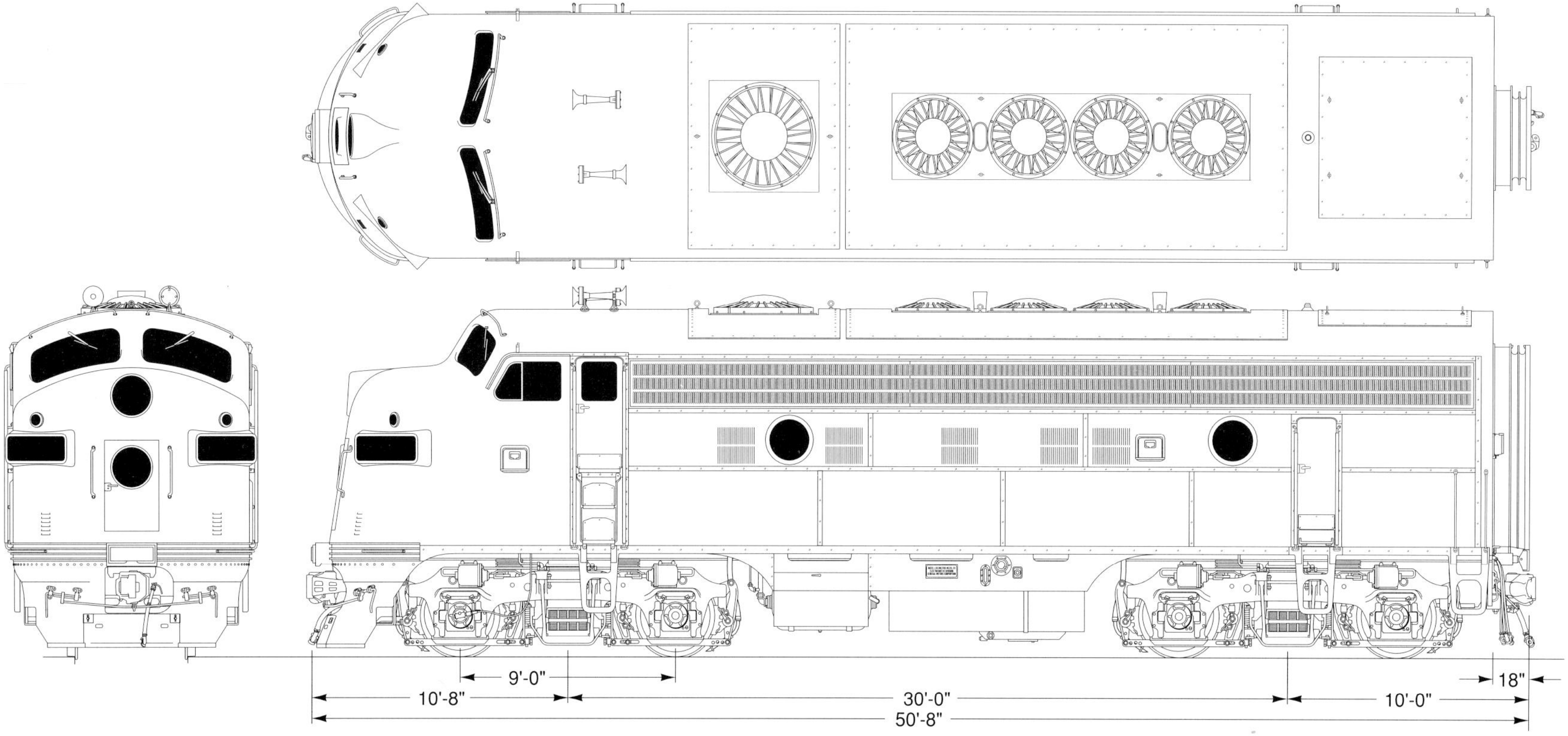

National Railways of Mexico F9 No. 7009, built in 1954. Electro-Motive photo.

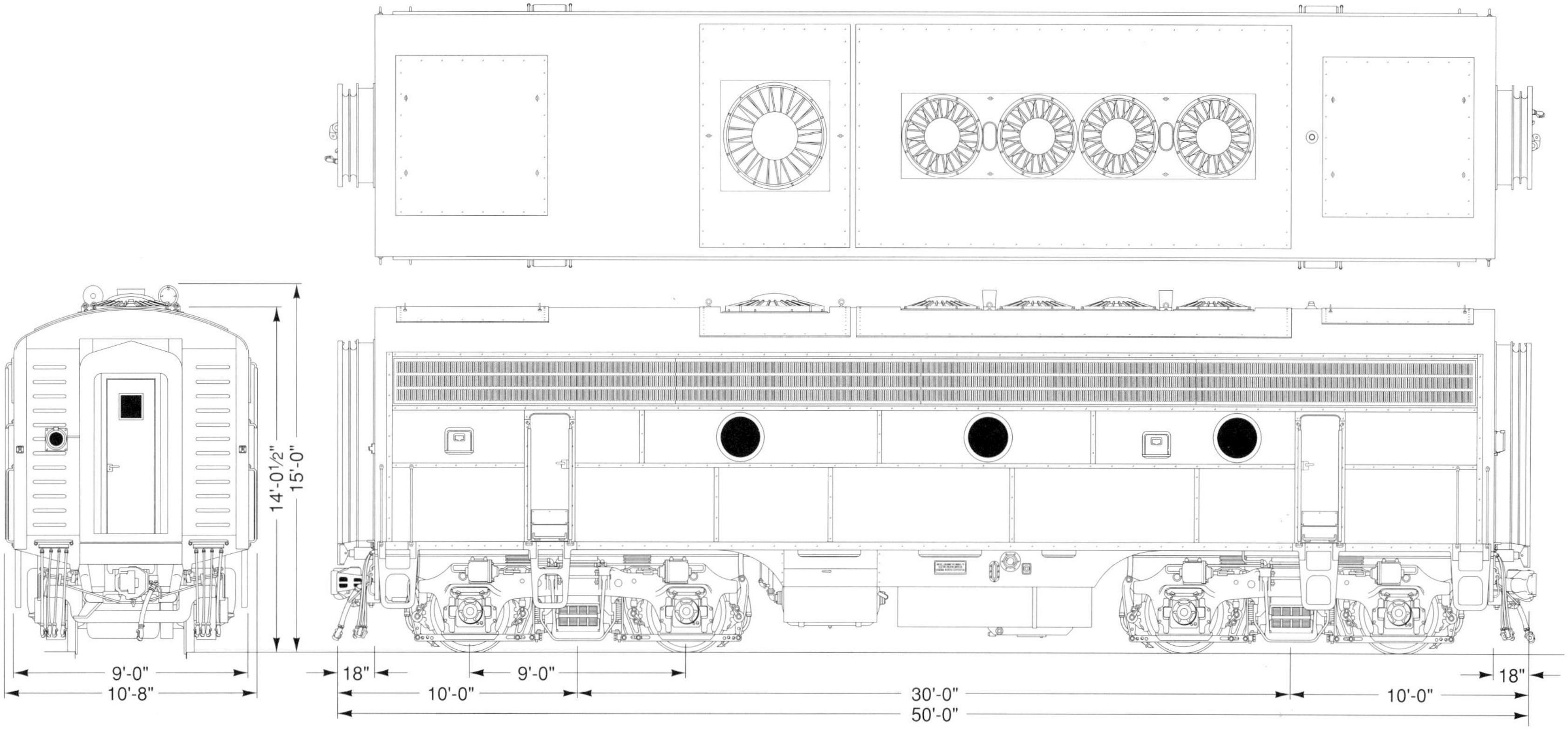

The FP7 was a passenger version of the F7, stretched four feet longer to accommodate a higher-capacity steam generator with a larger water tank. This extra length is apparent behind the forward porthole. A total of 372 FP7s were built from June 1949 to December 1953. No FP B units were built.

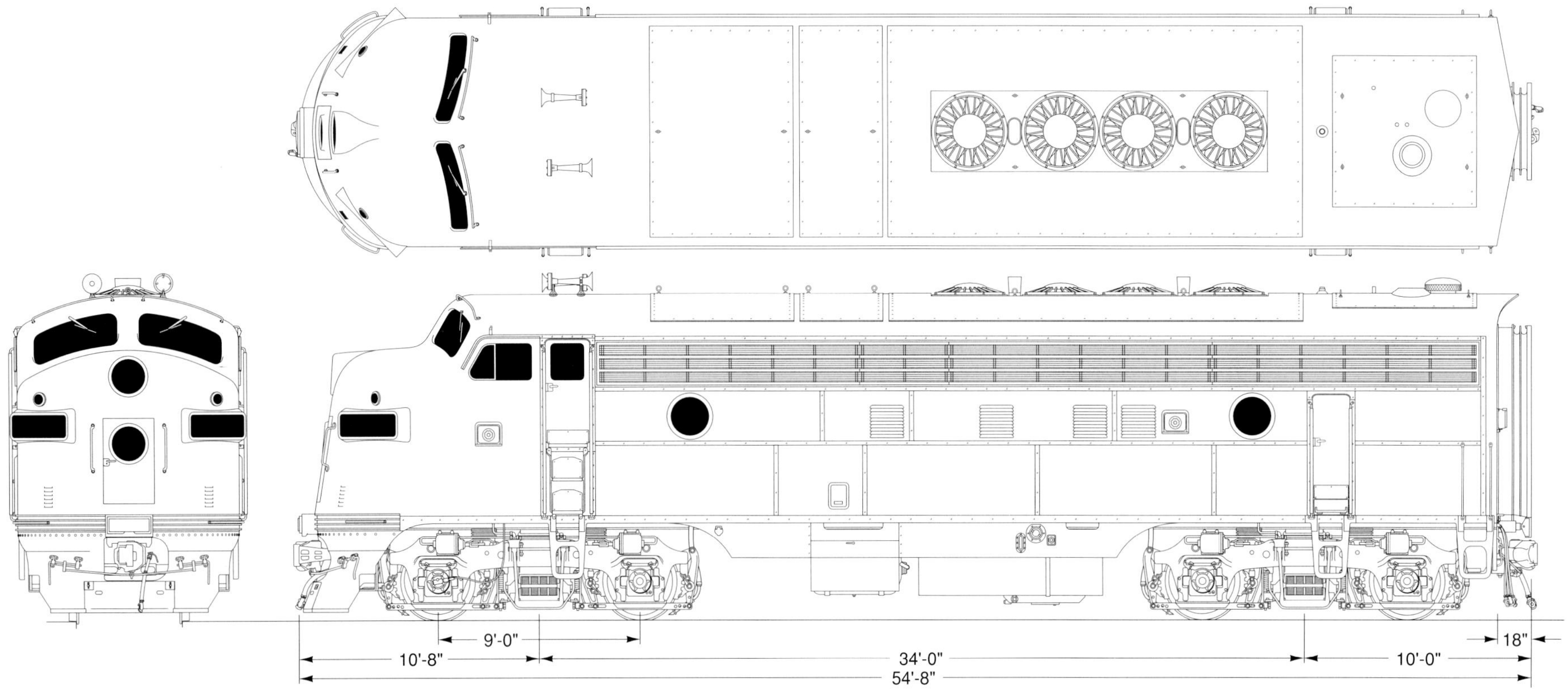

Reading FP7 No. 901, built in May 1950. Electro-Motive photo.

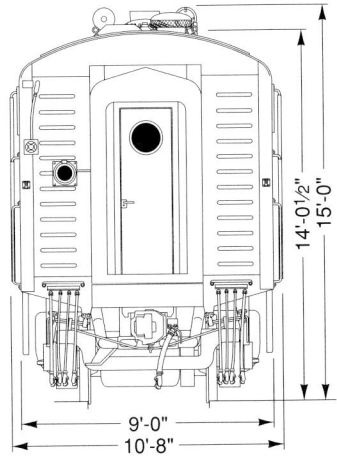

The FP9 was purchased only by Canadian National, Canadian Pacific, National Railways of Mexico, and Saudi Arabia. Canadian National No. 6501 was built at GMD in 1954. A total of 86 FP9s were built from February 1954 to July 1958.

Canadian National FP9 No. 6501, built in 1954. Photo by Larry Russell; collection of Louis A. Marre.

FL9

The 1,800-hp FL9 was built at the request of the New York, New Haven & Hartford. The railroad wanted a diesel locomotive that could also be underground in third-rail electrified territory into Grand Central Station. The FL9 had a third-rail pickup shoe mounted on each side of its six-wheel Flexicoil rear trucks, which were unique to this model. Sixty were built: 30 1,750-hp units in 1957 and 30 locomotives from September through November 1960. These were the last F units built.

New Haven FL9 No. 2001. Electro-Motive photo.

F Units Hard at Work

▶ They weren't good for switching, but it was tough to beat the view forward from an F unit. Many engine crews likened the ride on an F unit to that of a Cadillac, with a panoramic view high above the track. Ted Benson photo.

◀ Santa Fe passenger FT No. 166 and F3 32, both wearing the railroad's famous red-and-silver passenger warbonnet scheme, rest at Los Angeles Union Station, where they have just arrived pulling the *Chief* and *Super Chief*. That's Los Angeles City Hall at the front of 166. Santa Fe Railway photo.

◀ The *Fast Mail*—the first section of Santa Fe train No. 7—pauses to discharge mail and express at San Bernardino, California, behind passenger F3 No. 23. The full moon of this evening in 1951 illuminates San Gorgonio Peak in the distance, which marks the location of Beaumont Pass. Photo by Richard Steinheimer.

▲ Four F7s, wearing the Santa Fe's blue and yellow freight scheme, back toward their train at Barstow, California. Number 226 was built in December 1949. Santa Fe Railway photo by R. Collins Bradley.

▶ The engineer of Santa Fe Extra 235 West has this view of an F7-powered eastbound extra as his train drifts toward Ash Fork, Arizona. Photo by Wallace W. Abbey, April 1953.

▶▶ Santa Fe local No. 75, powered by passenger FT No. 415, is in the hole at Ponto, California, as local No. 83, a pair of Budd Rail Diesel Cars, passes it. The FT-powered train took four hours and 15 minutes to travel from San Diego to Los Angeles, while train 83, which doesn't have as much head-end work to do, will make the trip in 2:45. Photo by Richard Steinheimer, December 1952.

◀ An A-B-B-A set of Santa Fe F7s in the blue and yellow whiskers scheme, with white extra flags flying, hauls a long string of reefers eastbound between Ashfork and Williams, Arizona. Number 206 was built in July 1949. Santa Fe Railway photo.

▲ Atlanta & St. Andrews Bay tied for the distinction of having the smallest F unit roster: one unit. The railroad bought No. 1501, a steam generator-equipped early Phase II F3, in June 1947. It's shown here on train No. 4 at Panama City, Georgia. Photo by M. B. Cooke.

▶ Crew members of an F7-powered Atlantic Coast Line freight look back for the signal to proceed southbound out of the Florence, South Carolina, yard on the evening of August 27, 1957. Photo by William D. Middleton.

◀ Mid-afternoon at the Wilmington, North Carolina, station finds a pair of Atlantic Coast Line FP7s on the point of train No. 42. The station crew is readying the six-year-old purple-and-silver units for the train's 7 p.m. departure to Rocky Mount, N. C., on August 27, 1957. An E7 waits with its train on the next track. Photo by William D. Middleton.

◀◀ The Baltimore & Ohio owned 355 F units, including four-unit F7 set No. 186, shown here in 1950. The blue, gray, and black diesels were built in September 1949. Baltimore & Ohio photo.

◀ Baltimore & Ohio F7 No. 275 pauses at KG tower at Point-of-Rocks, Maryland. The locomotive is running light (by itself with no train), bound for the engine terminal at Brunswick, Md. The B&O's F units had cast heralds on their nose doors. Photo by R. Clarke Jr.

▲ Early FTs, such as Baltimore & Ohio No. 1 (built in August 1942), had dynamic brake enclosures with angular side fairings. This locomotive was soon renumbered 101. Photo by R. H. Payne; collection of Louis A. Marre and Gordon B. Mott.

◀ Bangor & Aroostook No. 44, a late Phase II F3 built in May 1948, arrives in Brownville Junction, Maine, with the Paper Train from Millinocket. Photo by Jim Shaughnessy.

▲ It's snowing on March 4, 1955, as Bangor & Aroostook F3 No. 42 arrives in West Seboois, Maine, with a string of empty cars bound for potato loading. The early Phase II F3, built in October 1947, is teamed up with a GP7 and another F3. Photo by Allen A. Sharp.

▶ Bessemer & Lake Erie owned a total of 54 F units: 28 F7As and 26 F7Bs built from 1950 to 1953. A matched four-unit set of the orange and black locomotives heads a string of Duluth, Missabe & Iron Range ore cars on January 1, 1951. Bessemer & Lake Erie photo.

▲ Boston & Maine A-B FT set No. 4203 was just 11 months old in this September 1944 view. The B&M's F units were maroon with yellow striping. Photo by H. P. Harvey.

▶ Canadian National owned 43 FP9s—half of the total number built. Here No. 6500 and F9B No. 6600 show off the CN's attractive green, yellow and black passenger scheme as they haul the *Ocean Limited* in 1955. Winterization hatches (the boxes atop the rear radiator fans) redirected warm air from the radiator fan back into the engine compartment in cold weather. Canadian National photo.

▶▶ An A-B set of late Phase II F3s in the Canadian National green and yellow freight scheme do some switching after arriving in New London, Connecticut, with train No. 490 from Sarnia, Ontario, in 1952. The locomotives were built in May 1948. Note the three-chime horn in place of the standard pair of single-note horns. Rail Photo Service photo by G. C. Corey.

◀ A Canadian Pacific FP7 and F7B pose for the company photographer in this early 1950s scene taken in the Canadian Rockies. The Canadian-built (GMD) diesels were delivered in 1952. Canadian Pacific photo.

▲ The brakeman aboard Canadian Pacific FP7 No. 4028 gets ready to climb down to line up the switches for a meet. The maroon and gray locomotives are leading train No. 13, the *Mountaineer*, at Banff, Alberta, on July 2, 1952. Photo by John C. Illman.

▲ Train No. 2 arrives at Banff, Alberta, on July 2, 1952, behind FP7 No. 4062. The yard engine, 2-8-2 No. 5248, is getting ready to add some passenger cars to the consist of No. 2. Photo by John C. Illman.

▶ The Chicago & Eastern Illinois rostered 16 F3 A units and 7 B units, all painted blue and yellow. Here Phase IV F3 No. 1408, built in December 1948, leads a train in early 1951. Chicago & Eastern Illinois photo.

◀ An A-B-A set of Chicago & North Western F7s rolls a freight across the wide-open prairie for which the railroad is known. David P. Morgan library collection.

◀◀ Chicago & North Western F7As 4094-C and 4094-A lead an iron ore train at Wakefield, Michigan, en route to Escanaba, Mich., on August 12, 1951. The year-old Fs have the optional lower headlight and are painted green, yellow, and black. Photo by E. Treloar.

◀ Several North Western F units await servicing at 40th Street in Chicago. Number 4071-C is an F7 built in March 1949. Chicago & North Western photo.

▶ Chicago, Burlington & Quincy No. 101 was one of 16 four-unit FT sets allotted the railroad by the War Production Board in 1944. The engineer strikes up a pose for the company photographer on February 2, 1944. Chicago, Burlington & Quincy photo.

▶▶ Four Chicago, Burlington & Quincy FTs lead a freight down Mason Street at Fort Collins, Colorado, on subsidiary Colorado & Southern. The F received a lower headlight in the late 1940s. Photo by Fred H. Matthews, Jr., June 25, 1951.

◀ Maroon and red F7 No. 153 and a sister A unit arrive in Minneapolis, Minnesota, with the *Mill Cities Limited* on August 4, 1950. The CGW had nine F units equipped with steam generators for passenger service: four F7s, three F3s, and two FP7s. Photo by Bob Borcherding.

▲ Midwestern granger railroad Chicago Great Western didn't have the most F units, but the railroad was known for stringing the greatest number of them together at one time. Here seven Fs, led by late Phase II F3 No. 107A, lug their 11,000-ton Chicago-bound train out of Oelwein, Iowa, in 1952. Chicago Great Western photo.

▶ Milwaukee Road F3 No. 82 was one of eight A-B sets delivered to the railroad in February 1949. The two-tone gray engine has the optional passenger-style pilot. Electro-Motive photo.

▶▶ Maroon and orange FP7 No. 93C, one of 32 owned by the Milwaukee Road, leads a passenger train through Canton, South Dakota, in the summer of 1950. An Alco RSC-2 switches some passenger cars in the background. Photo by Henry J. McCord.

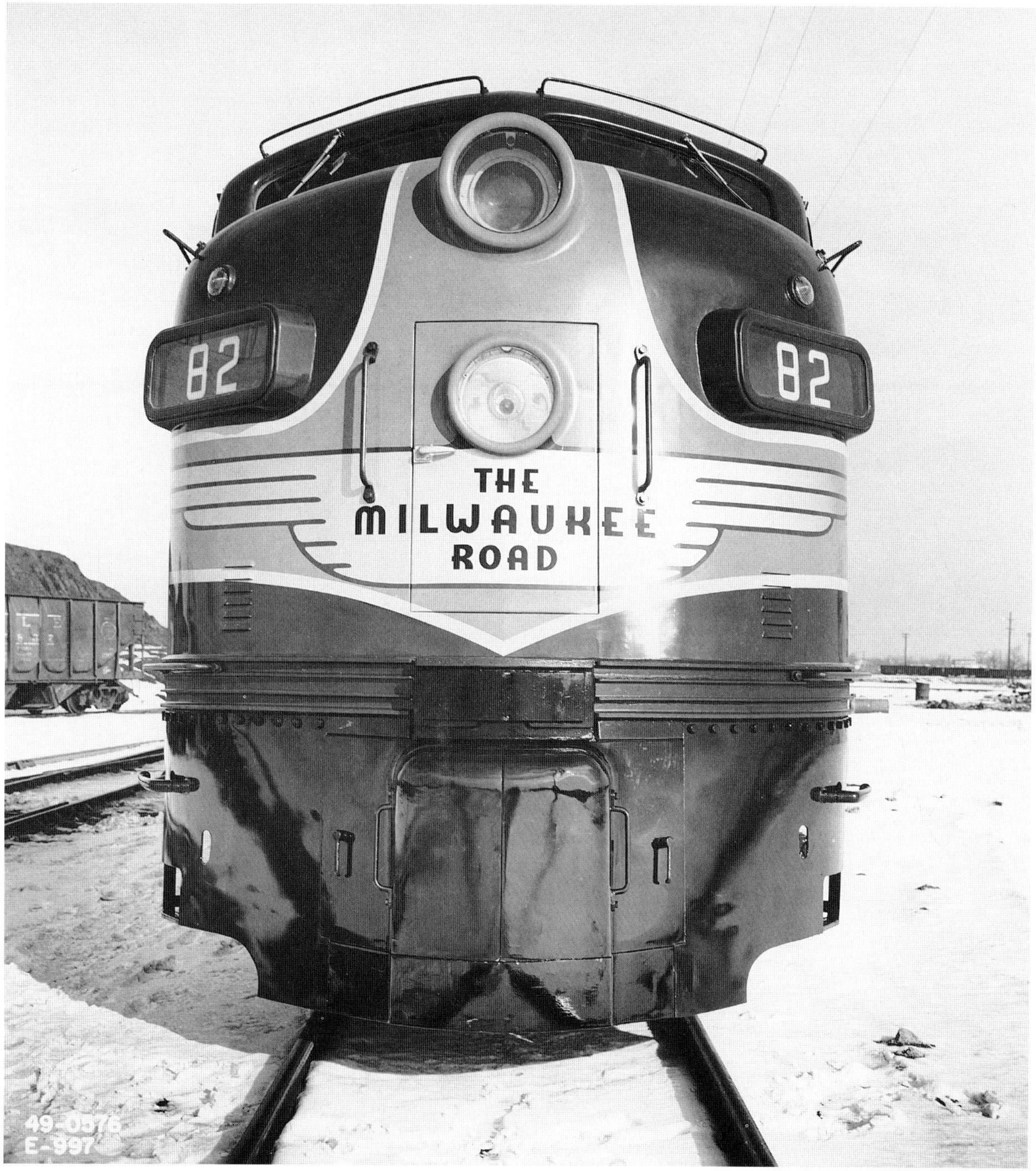

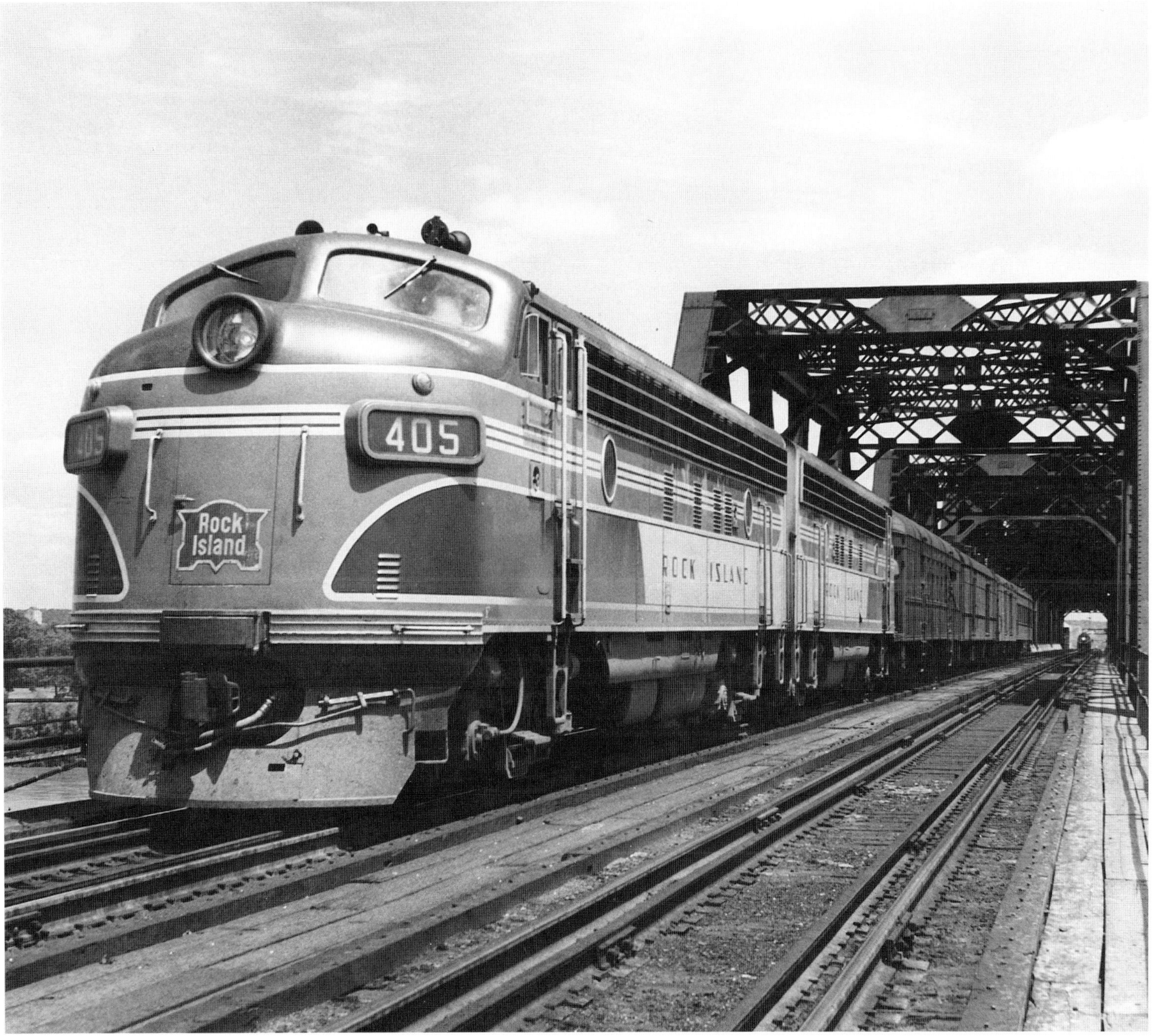

◂◂ Brand-new Milwaukee Road F7 No. 70A leads two sister F units on a freight at La Grange, Illinois, in August 1950. Photo by Robert Milner.

◂ Number 405 and a sister FP7 lead Rock Island train No. 5, the *Des Moines-Omaha Limited,* across the Missouri River into Omaha, Nebraska, on June 19, 1950. Photo by Donald Sims.

▶ The Rock Island's ten passenger-service FP7s, built in June and July 1949, were among the first FP7s built. They wore the railroad's silver, red, and black passenger scheme. Here No. 403 leads a Chicago suburban train out of Joliet, Illinois. Photo by Wallace W. Abbey.

▶▶ A trio of Chicago, Rock Island & Pacific F7s rests on a diesel servicing track at the railroad's Silvis, Illinois, yard in May 1952. The Rock Island painted its freight-service Fs in red and black, and in the early 1950s they wore the Rocket Freight "RF" logo. Photo by Charles H. Kerrigan.

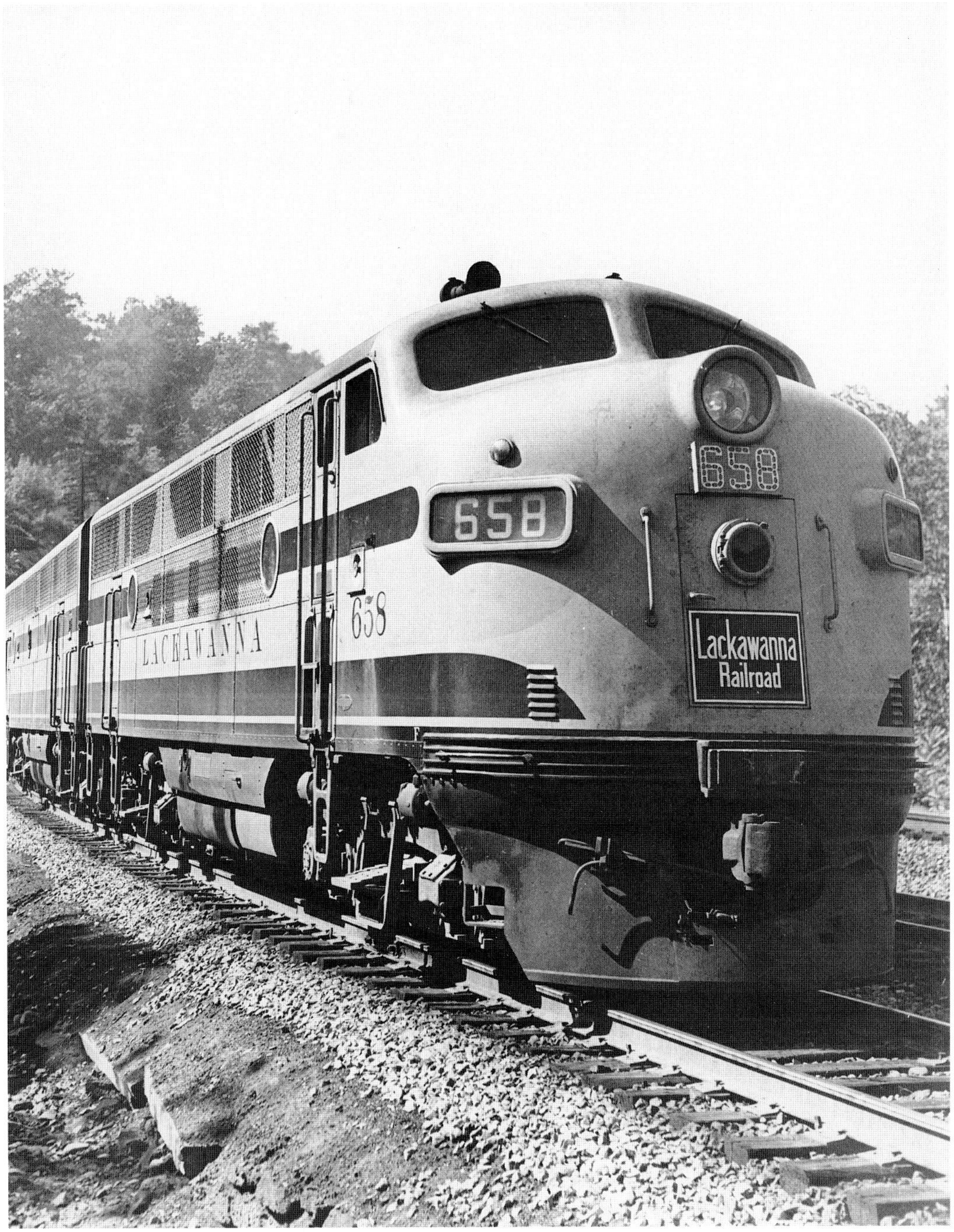

◀◀ Gray and yellow Clinchfield F7 No. 814 leads time freight No. 95 out of Skaggs Tunnel and across Russell Fork Bridge in the summer of 1952. Rail Photo Service photo by B. F. Cutler.

◀ Delaware, Lackawanna & Western F3 No. 658 shows the railroad's freight scheme, which featured the same colors as passenger units, but in a different pattern. These Fs are on a freight train near Scranton, Pennsylvania, in 1951. The light on the nose door of some Lackawanna Fs was red, used on locomotives at the rear of trains in helper service. Photo by S. Botsko.

▶ The Delaware, Lackawanna & Western had several Fs equipped with steam generators for passenger service. Here late Phase II F3 No. 803C and an earlier Phase I F3 lead the *Merchant's Express* eastbound at Dover, New Jersey, on December 29, 1951. The Lackawanna's Fs were gray with maroon bands and yellow noses and stripes. Photo by H. C. McClain.

▶▶ Denver & Rio Grande Western No. 540 was one of three four-unit sets of FTs purchased for freight service in January and February 1942. The railroad later acquired another nine sets of FTs. The black and yellow locomotives were eventually given individual numbers in the 5400 series. Denver & Rio Grande Western photo.

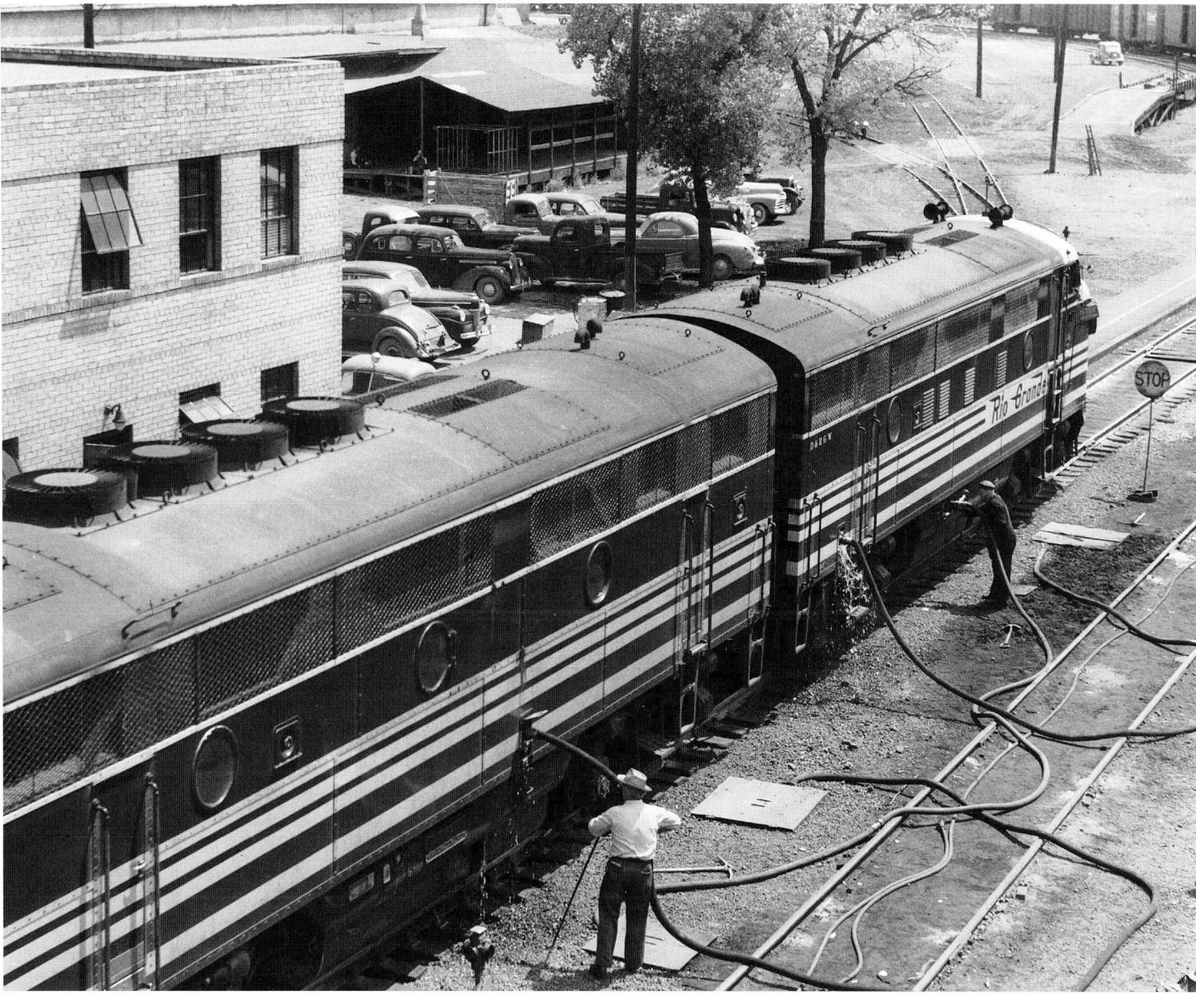

◀◀ The Rio Grande used F units on passenger trains over its rugged Rocky Mountain line. You can see the steam generator stacks and vents on the rear of each B unit roof. Here F3 No. 5521 climbs South Boulder Canyon near Pinecliff, Colorado, with the *California Zephyr.* Photo by R. H. Kindig, April 28, 1951.

◀ Pausing with the westbound *California Zephyr,* F3 set No. 5521 takes on water and fuel at Grand Junction, Colorado, on August 24, 1951. Photo by E. E. Law Jr.

▶ Electro-Motive F9 demonstrator No. 975 is open for inspection in the mid-1950s. Note the multiple-unit (m.u.) hatch between the headlight and number board. Electro-Motive photo.

▶▶ Matched Erie F3 set No. 708-D, -C, -B, and -A heads a fast freight across the Chemung River at Elmira, New York, in early 1957. The black and yellow early Phase II F3s were built in November 1947. Photo by Jim Shaughnessy.

◀◀ Among the last standard F units built were five F9As and six F9Bs for Northern Minnesota's Erie Mining Co. Here four of the F9s, painted yellow with maroon stripes, ease 100 ore loads down a 2 percent grade. David P. Morgan Library collection.

◀ Several new yellow and red Florida East Coast Phase IV F3s line up for a publicity shot in front of the Dade County Courthouse in Miami, Florida, in January 1949. Florida East Coast photo by Harry M. Wolfe.

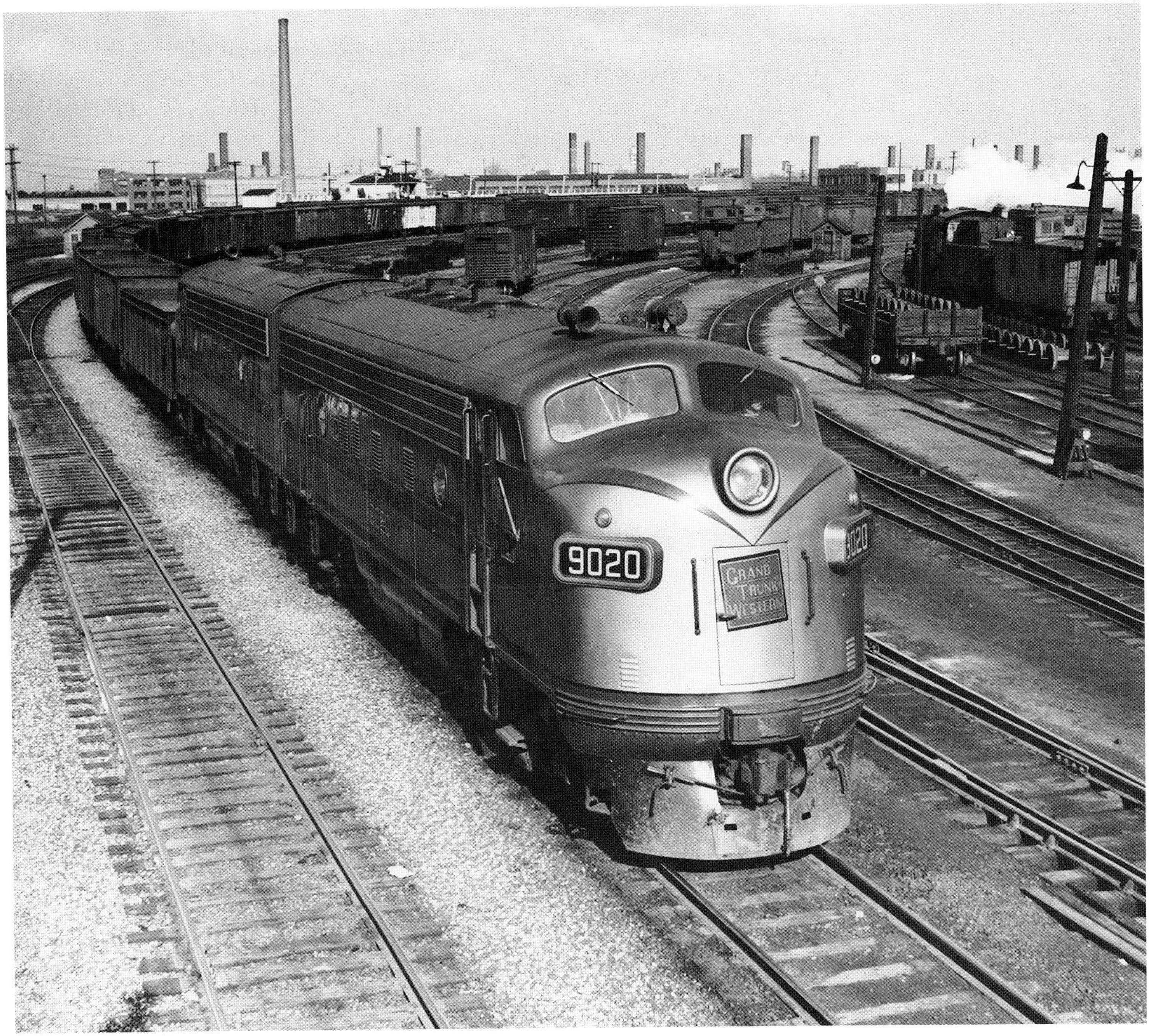

◄◄ The Georgia Railroad operated just four F units: an F3 and three FP7s, including No. 1004, built in February 1950. The coaling tower at Social Circle, Georgia, was no longer needed by the time of this September 13, 1951, photo. Photo by William J. Husa Jr.

◄ Among the flashiest F unit paint schemes was the gold and green of Canadian National-controlled Grand Trunk Western. Here Phase IV F3 No. 9020, built in September 1948, and a sister locomotive lead a freight train. Canadian National photo.

◀◀ All the details of late Phase II F7s are apparent on this view of a trio of Great Northern units: large (48-inch) dynamic brake fan, horizontal-slit air intake louvers, and Farr-Air side grills. The three units, built in 1952, are heading a 180-car iron ore train. Great Northern Railway photo.

◀ The fireman aboard Great Northern FT No. 418D waves to a fisherman in a classic publicity shot. The GN had 95 FTs, second only to Santa Fe. No. 418 was built in August 1944. Great Northern Railway photo.

◀ Workers clean the windshields and headlights of Great Northern Phase II F3 No. 358 as the westbound *Fast Mail* pauses at the Spokane, Washington, depot on July 11, 1948. These orange and green Fs, built in December 1947, were equipped with steam generators and had passenger-style pilots. Photo by Wade J. Stevenson.

▲ A trio of Gulf, Mobile & Ohio Phase I F3s rolls a freight train south of Bloomington, Illinois, on September 10, 1947. Number 800A was built in December 1946. The locomotives wore an attractive two-tone red paint scheme. Photo by Paul Stringham.

▶ Two Gulf, Mobile & Ohio early Phase II F3s are in charge of the *Ann Rutledge* as it leaves Pennsylvania Railroad trackage after crossing the lift bridge at 21st Street in Chicago on July 20, 1948. The railroad owned nine F3As and three F3Bs with steam generators for passenger service. Photo by Robert H. Milner.

▶ Kansas City Southern train No. 77 leans into a curve on Rich Mountain, between Page, Oklahoma, and Howard, Arkansas, behind four F units on January 17, 1953. Lead locomotive No. 31 is a Phase IV F3 built in November 1947. Photo by A. E. Brown.

◄◄ Four fresh-from-the-factory red, black, and yellow Kansas City Southern F3s power an eastbound Chicago Great Western freight train at St. Charles, Illinois, in March 1948. Photo by Henry McCord.

◄ Lehigh Valley's F units wore a deep red scheme with black stripes. Here four-year-old Phase IV F3 No. 526 leads a freight near Pittston, Pennsylvania, on May 19, 1952. Photo by William L. Farber.

◀◀ Two black and yellow Louisville & Nashville F7s lead 106 cars south of Montgomery, Alabama, on May 26, 1955. Photo by J. Parker Lamb.

◀ The fireman of a well-scrubbed Louisville & Nashville F7 glances at a rust-coated steam locomotive in 1955. Photo by William A. Akin.

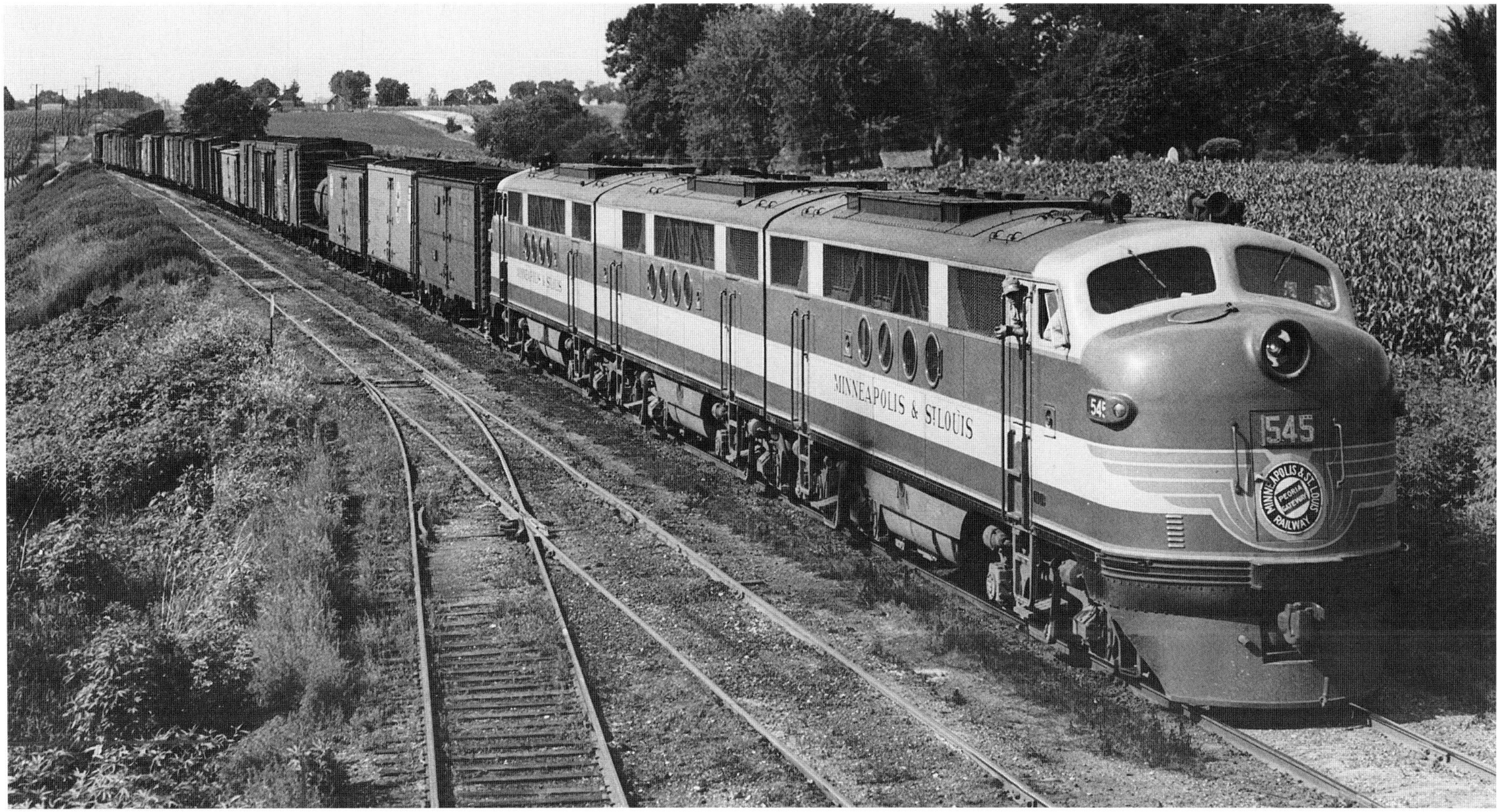

◀ Looking a lot like a perfectly matched model train set, Louisville & Nashville F7 No. 903 has a single L&N boxcar and caboose in tow at El Dorado, Illinois, on July 6, 1959. Photo by J. Parker Lamb.

▲ The Minneapolis & St. Louis bought two A-B-A sets of FTs in April 1945. The B units of the three-unit sets were shorter than standard FT B units and dubbed FTSBs. The green and yellow locomotives head a Peoria-bound freight at Oscaloosa, Iowa, on July 28, 1950. Photo by Robert Milner.

▲ Two Phase II F3s lead M&StL time freight No. 90 through Oscaloosa, Iowa, on July 28, 1950. Note the difference in paint schemes between the March 1948 F3s and the earlier FTs on page 87. Photo by Robert Milner.

▶ Missouri-Kansas-Texas Extra 203A South leaves Fort Worth, Texas, behind three early Phase II F3s on October 19, 1952. The locomotives, built in June 1947, were red with aluminum lower sides and yellow noses. Photo by R. S. Plummer.

▶ The fireman of Missouri Pacific eastbound manifest freight No. 83 leans out to pick up orders at Rusk, Texas, in August 1951. The locomotive is F3 No. 535. Photo by Donald Sims.

▶▶ Missouri Pacific's FTs didn't have dynamic brakes, as evidenced by the clean roof lines on A-B set No. 508. The blue and light gray locomotives were brand new when photographed at Dupo, Illinois, on October 27, 1944. Collection of Louis A. Marre.

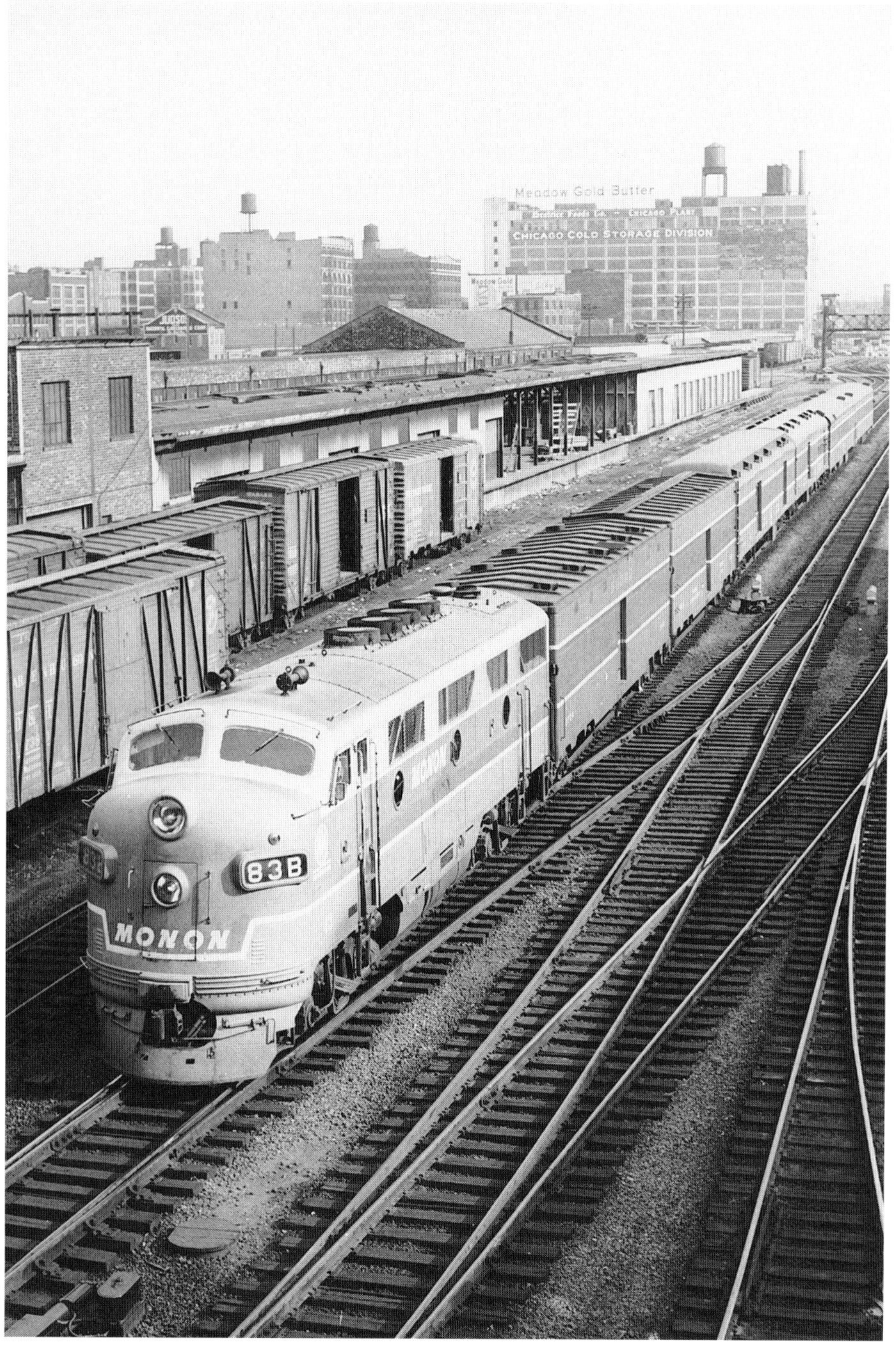

▲ Monon's freight-service Fs, such as Phase I F3 No. 63, were painted gold and black. David P. Morgan Library collection.

▶ A Monon local passenger train arrives in Chicago behind Phase I F3 No. 83B in the early 1950s. Monon's ten passenger-service Fs were painted red and gray. Photo by George Zygmund.

▲ The National Railways of Mexico operated 128 F units, including 28 F2s (14 As and 14Bs), such as No. 6203, shown at San Luis Potosi on December 5, 1948. Photo by George W. Sisk; collection of Louis A. Marre.

▶ The sun shines off the nose of New York Central F3 No. 1877 on March 9, 1961. Note the cast herald on the nose door. Collection of Louis A. Marre.

▶▶ The lineup at the New York Central's Collinwood, Ohio, yard in 1952 included F7 No. 1648, F3 No. 1613, F7 No. 1688, and F3 No. 1615. All were painted in the Central's black and gray lightning stripe freight scheme. New York Central photo.

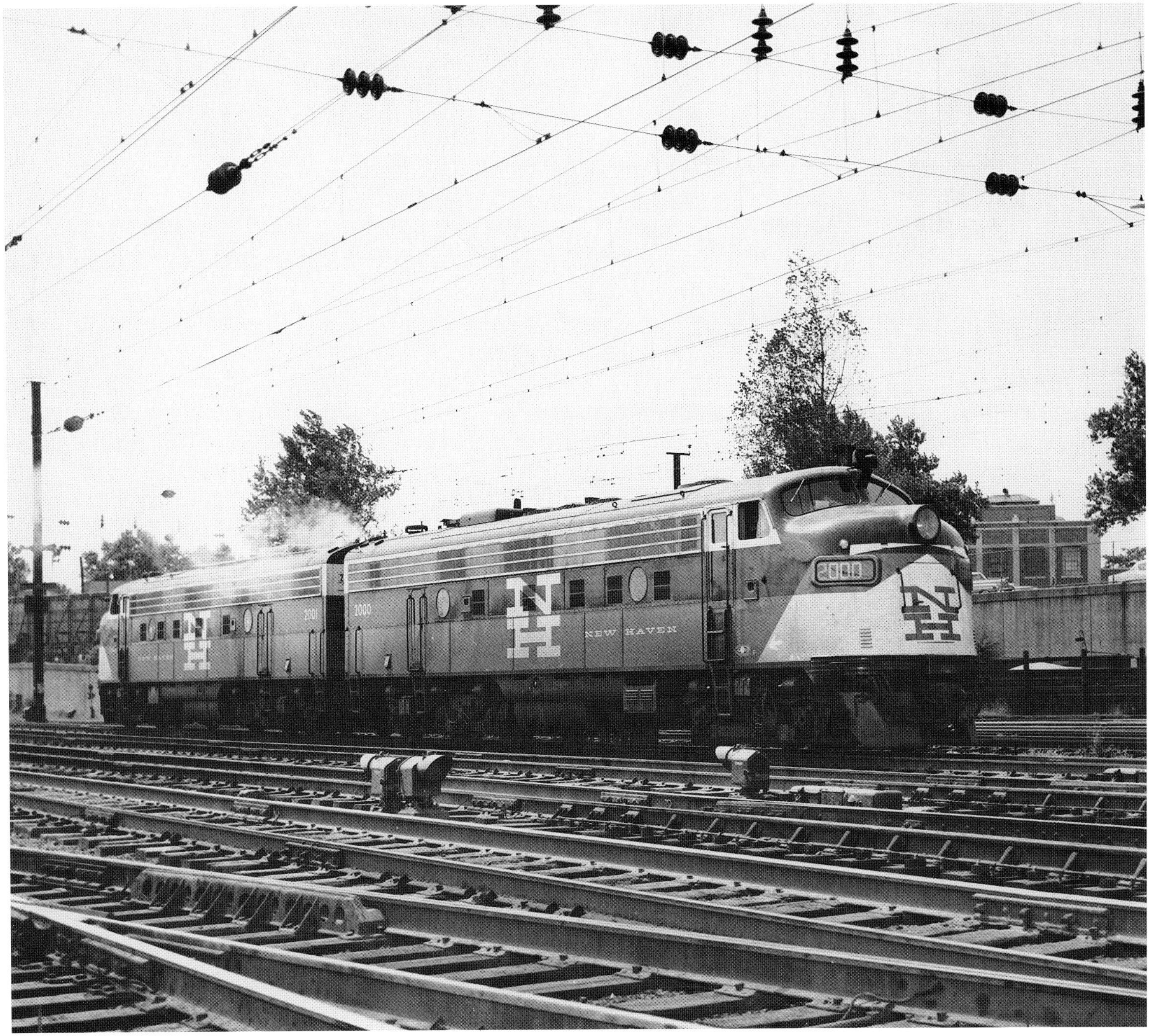

◀◀ New York, New Haven & Hartford train No. 54 pops into daylight at East 97th Street in New York after emerging from the Park Avenue Tunnel behind two FL9s. The New Haven's red, black, and white Fs had Hancock horns mounted above the windshields. Photo by Don Wood, February 15, 1958.

◀ The first two FL9s built, New Haven Nos. 2000 and 2001, rest under a maze of catenary in 1960. The engines were delivered in the orange, black, and white McGinnis scheme. New York, New Haven & Hartford photo.

◀ The fireman of a New York, Ontario & Western F3 looks back on his train at Cadosia, New York, on June 20, 1948. Number 502 was built in February 1948. Photo by Charles S. Small.

▲ The New York, Ontario & Western began dieselizing with 18 FTs (9 As and 9 Bs) built in May 1945. Here No. 806 leads a freight bound for Maybrook, New York, at Campbell Hall Tower in 1948. The engines were gray, yellow, and orange. Rail Photo Service photo by James D. Bennett.

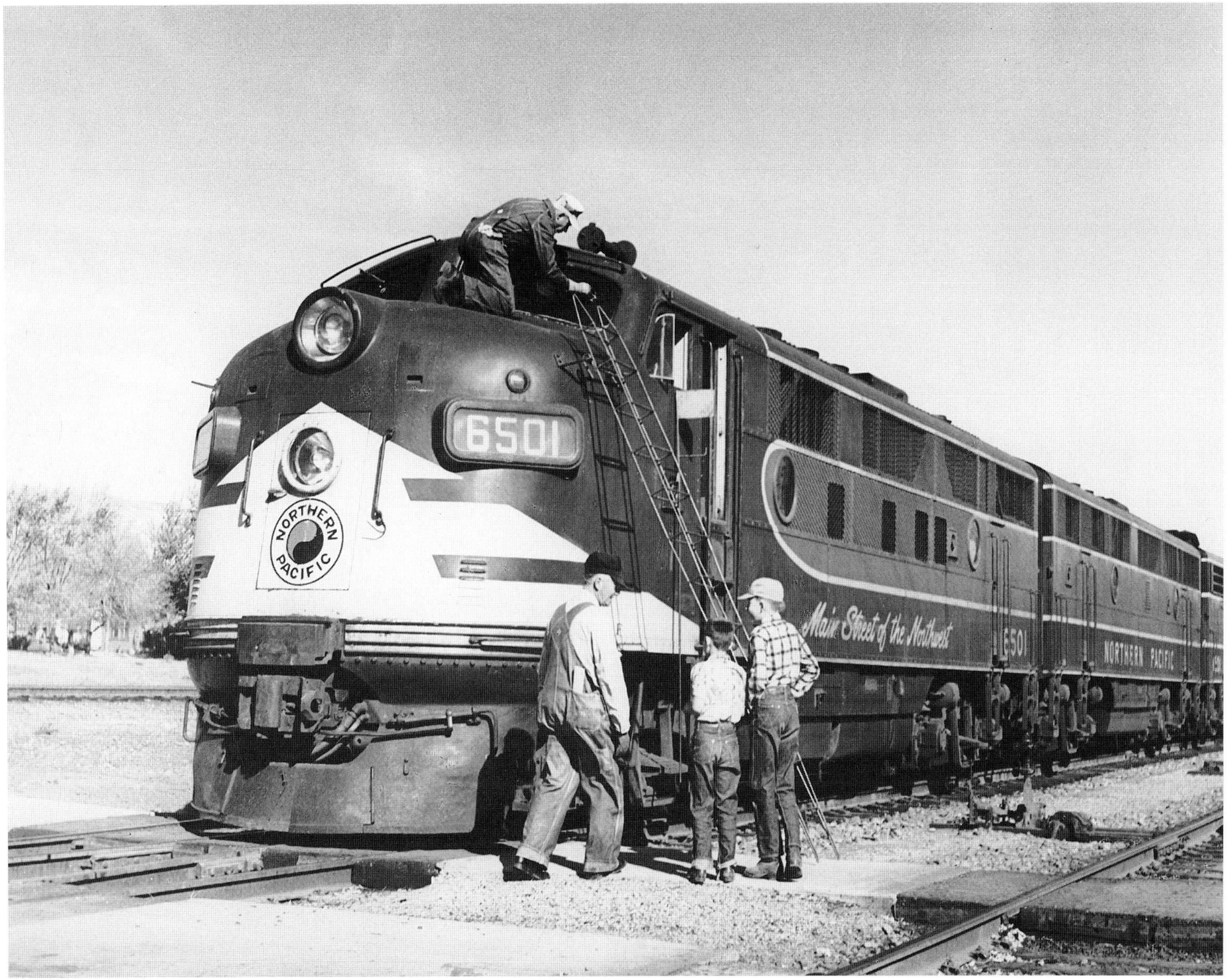

◀◀ The first storm of winter is blowing in as three students walk in front of Northern Pacific F3 No. 6504, which is pausing for a station stop. The two-tone green diesel, painted in the 1953 Loewy passenger scheme, is heading train No. 3, the St. Paul, Minnesota, to Glendive, Montana, mail train in November 1957. Photo by Richard Steinheimer.

◀ Lunchtime at a local school in Livingston, Montana, coincided with the arrival of the westbound *North Coast Limited,* meaning that servicing crews always had plenty of supervision. After dieselization the Northern Pacific relied exclusively on F units to power its passenger trains, and early Phase II F3 No. 6501, built in January 1947, is painted in the NP's early two-tone green and yellow passenger scheme. Photo by Donald Sims.

◂ Westbound Northern Pacific hotshot freight No. 603 pauses at Jamestown, North Dakota, an hour late because of a snowstorm. On the point is one of the NP's F3s wearing the black and yellow freight scheme. Photo by Richard Steinheimer, November 1957.

▲ Two Pennsylvania Railroad F3s and an F7 grind up Pennsylvania's Aligrappas Gorge with a manifest freight in October 1949. The railings atop the A units are antennas for the Pennsy's induction (radio) train telephones. Photo by Fred McLeod.

▲ Pennsy F3 No. 9513, amid a cloud of exhaust, pops out of the Rockville, Pennsylvania, overpass with an Altoona-bound freight in October 1954. Note the single air horn and rectangular dynamic brake fan openings on the roof. Photo by Don Wood.

▲ A Reading way freight led by F7 No. 266 has just set out cars at the St. Nicholas, Pennsylvania, yard on its way to Philadelphia. The Reading's Fs were black with green and yellow stripes. Photo by S. Botsko.

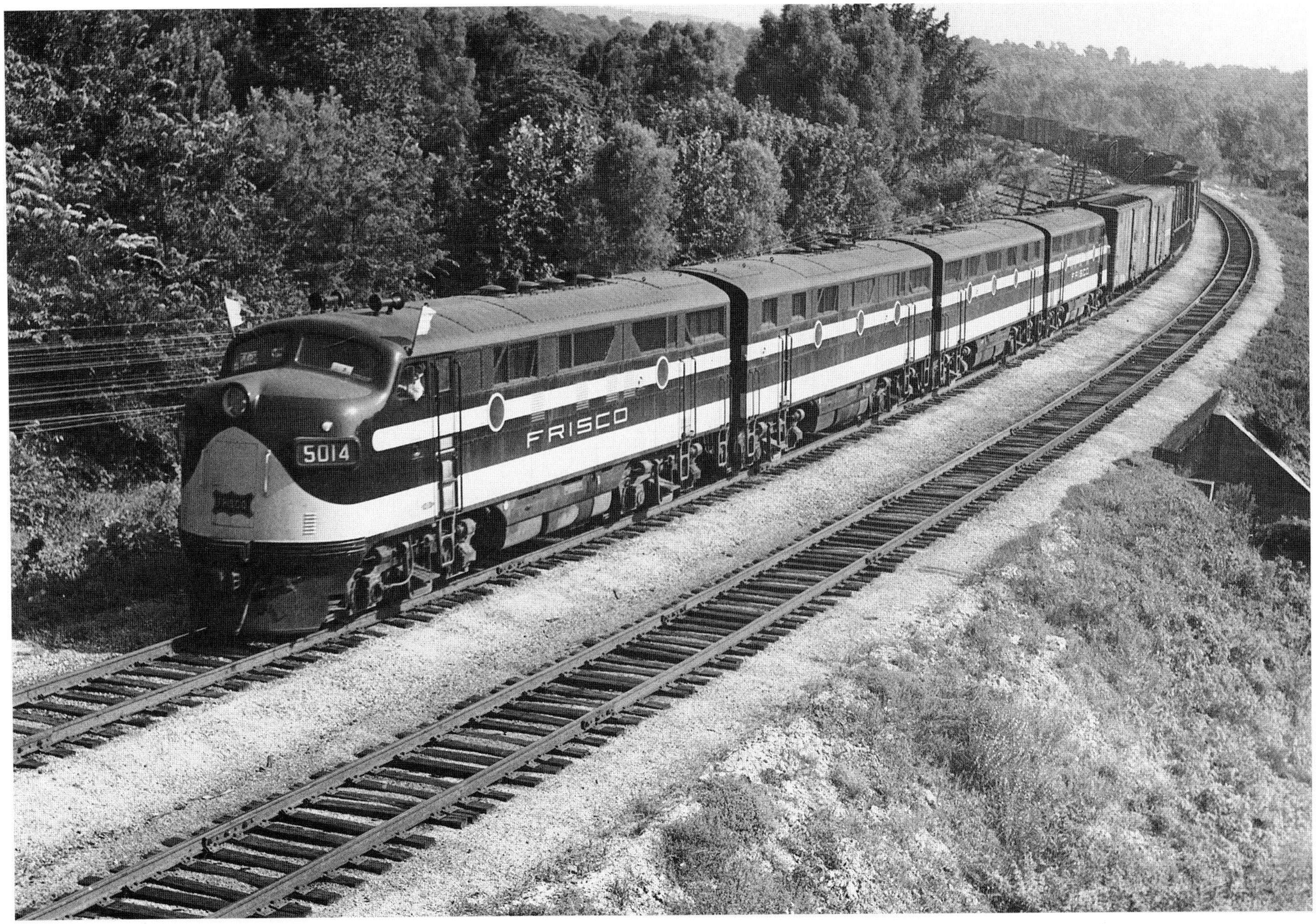

◀ With white extra flags flying in the wind, St. Louis-San Francisco Phase III F3 No. 5014, built in June 1948, heads a freight just after delivery. The Frisco's Fs were painted black and white, with no lettering on the B units. St. Louis-San Francisco Railway photo.

▲ St. Louis Southwestern (Cotton Belt) FT No. 910 shows off the gray paint scheme initially applied to the railroad's F units. The railroad owned 20 FTs built in 1944 and 1945. Photo by R. J. Foster, 1949.

▲ Number 963, a Phase II F7, wears the black widow scheme of parent Southern Pacific, but with St. Louis Southwestern initials along the side and a Cotton Belt herald on the nose. Photo by R. S. Plummer, May 25, 1952.

▶ Workers scrub down brand-new Soo Line (Wisconsin Central) FP7 No. 2500 on a foggy morning in Duluth, Minnesota. The maroon and Dulux Gold engine had just arrived with train 17 from Chicago on July 17, 1950. Photo by Robert Milner.

◀ Soo Line FP7 No. 501 rests at the Milwaukee Road station in Minneapolis on the evening of August 28, 1958. The locomotive and a B unit had arrived from St. Paul with train No. 19, which will soon head to Portal, North Dakota. Photo by William D. Middleton.

▲ Southern Railway F3 No. 4139 basks in the sun at Atlanta on October 20, 1947. The engine was built in December 1946. Photo by James G. LaVake.

▲ It's a hazy day as four Southern Railway Phase I F3s head across a tall steel trestle in Virginia, trailing a long string of tank cars. The Southern received the lead unit, No. 4147, in January 1947. Southern Railway photo.

⏪ A four-unit set of Southern Railway FTs (actually owned by subsidiary Cincinnati, New Orleans & Texas Pacific) leads a freight around a sweeping curved steel trestle near Burnside, Kentucky, in the mid-1940s. These black and light gray FTs, built in May 1941, aren't equipped with dynamic brakes. Southern Railway photo.

◀ Southern Pacific's classic black widow scheme featured a black body, silver nose, and orange and red striping. Here Phase IV F3 No. 6155, built in December 1948, leads the second section of train 803 near San Fernando, California, on March 19, 1950. Photo by Donald Sims.

◀ One of the Southern Pacific's Phase II F7s speeds along with train 371 in the mid-1950s. Richard Steinheimer photo.

◀ Rebuilt Southern Pacific F7 No. 6311 has just battled through an ice storm at Dunsmuir, California. Photo by Richard Steinheimer.

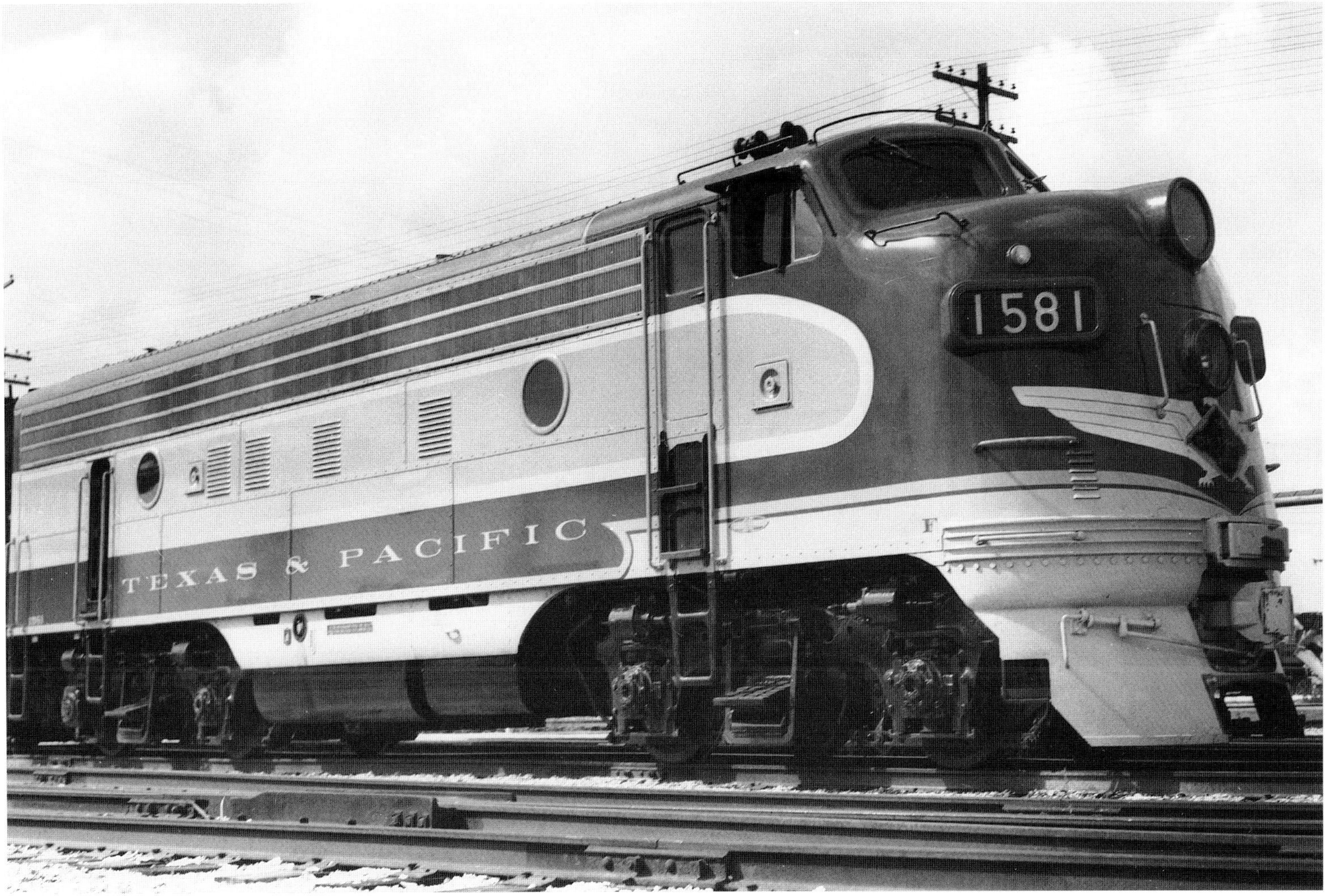

◀◀ A new crew comes on board Spokane, Portland & Seattle F3 No. 802 under the lights at Pasco, Washington. The F, built in November 1948, was one of three on the roster. It had just arrived with train No. 1. Photo by Donald Sims.

◀ Texas & Pacific F7 No. 1581 wears the full blue and gray *Eagle* livery of parent Missouri Pacific. The two-year old F was photographed at New Orleans in May 1954. Collection of Louis A. Marre.

▲ Union Pacific early Phase II F3 No. 1419, built in January 1948, leads three sisters on a stock train descending Cajon Pass in November 1948. The UP's Fs are yellow with red lettering. Photo by Donald Sims.

◀ The Detroit skyline rises in the background as a trio of Wabash F7s leads an eastbound train out of Windsor, Ontario, in January 1960. The Fs are painted blue, gray, and white. Photo by Bruce R. Meyer.

◀ Two F7s and a Geep pause with train No. 89 at Tolono, Illinois, on March 29, 1958. The Alco PA in the foreground is heading the eastbound *Cannonball*. Photo by J. Parker Lamb.

▲ An A-B-B-A set of Western Maryland F7s rolls past the depot at Highfield, Maryland, on July 11, 1953. The three-year-old diesels, painted black with yellow lettering, sport twin five-chime horns. Photo by Warren L. Bain Jr.

▲ The Western Pacific bought 12 A-B-B-A sets of FTs between November 1941 and November 1944. Number 902, built in December 1941, wears an experimental orange, aluminum, and black scheme in 1946. Western Pacific photo.

▶ Three steam-generator-equipped late Phase II F3s lead the *California Zephyr* across Rock Creek Bridge in the Feather River Canyon in 1950. The Western Pacific acquired the silver and orange Fs in June 1947. Rail Photo Service photo.

F Units Today

Hundreds of F units survived in mainline service through the 1960s, and some even made it through the 1970s and into the 1980s. But by that time the F units' age and carbody design had relegated them to secondary and shortline service, and by the 1990s a railfan generally had to go to a museum or take a trip on a dinner train to see an F unit working.

However, there was an exception. Way off the beaten trail, up in northern Minnesota's Iron Range, a group of F units kept grinding out the miles in the service for which they were built. Erie Mining bought a group of F9s new in 1957 to haul solid trains of taconite 150 miles from Hoyt Lakes, Minn., to Taconite Harbor on Lake Superior, and 40 years later that's exactly what they were still doing for Erie successor LTV Steel.

Dedication

This book is dedicated to my Dad, Cliff, who always encouraged my interest in trains. He didn't know an F unit from a Geep, but no matter how much of my allowance I spent on books and model trains (I know he cringed at times) or how much of the basement I filled with a layout, he was always proud of what I did (and rather amazed that I actually found a way to earn a living from my hobby!). Thanks, Dad.

Index

Railroads
Atchison, Topeka & Santa Fe: 6, 29-34
Atlanta & St. Andrews Bay: 35
Atlantic Coast Line: 36, 37
Baltimore & Ohio: 17, 38-40
Bangor & Aroostook: 41
Bessimer & Lake Erie: 42
Boston & Maine: 43
Canadian National: 26, 44, 45
Canadian Pacific: 46-48
Chesapeake & Ohio: 20
Chicago & Eastern Illinois: 49
Chicago & North Western: 50-53
Chicago, Burlington & Quincy: 15, 19, 20, 54, 55
Chicago Great Western: 56-57
Chicago, Milwaukee, St. Paul & Pacific: 58-60
Chicago, Rock Island & Pacific: 61-63
Clinchfield: 64
Delaware, Lackawanna & Western: 65, 66
Denver & Rio Grande Western: 67-69
Electro-Motive: 4, 6, 16, 70
Erie RR: 71
Erie Mining Co.: 72, 125
Florida East Coast: 73
Georgia RR: 74
Grand Trunk Western: 75
Great Northern: 76-78
Gulf, Mobile & Ohio: 17, 79
Kansas City Southern: 80-82
Lehigh Valley: 83
Louisville & Nashville: 84-86
Minneapolis & St. Louis: 87-88
Missouri-Kansas-Texas: 89
Missouri Pacific: 18, 90, 91
Monon: 92
National Rys. of Mexico: 23, 93
New York Central: 94, 95
New York, New Haven & Hartford: 27, 96, 97
New York, Ontario & Western: 98, 99
Northern Pacific: 100-102
Pennsylvania RR: 103
Reading: 25, 104
St. Louis-San Francisco: 105
St. Louis Southwestern: 106
Soo Line: 107, 108
Southern Ry.: 109-110
Southern Pacific: 111-115
Spokane, Portland & Seattle: 116
Texas & Pacific: 117
Union Pacific: 118
Wabash: 119, 120
Western Maryland: 121
Western Pacific: 122, 123

Locomotive types
FT
  ATSF: 6, 29, 33
  B&O: 40
  B&M: 43
  CB&Q: 54, 55
  D&RGW: 67
  EMC: 4, 6
  GN: 77
  M&StL: 87
  MP: 91
  NYO&W: 99
  SLSF: 106
  Southern: 110
  WP: 122
F2
  CB&Q: 15
  NdeM: 93
F3
  ATSF: 29, 30
  A&StAB: 35
  B&O: 17
  BAR: 41
  CN: 45
  C&EI: 49
  CB&Q: 19
  CGW: 57
  CMStP&P: 58
  DL&W: 65, 66
  D&RGW: 68, 69
  Erie RR: 71
  EMD: 16
  FEC: 73
  GTW: 75
  GN: 78
  GM&O: 17, 79
  KCS: 81, 82
  LV: 83
  M&StL: 88
  MKT: 89
  MP: 18, 90
  Monon: 92
  NYC: 94, 95
  NYO&W: 98
  NP: 100, 101, 102
  PRR: 103
  SLSF: 105
  Southern: 109
  SP: 111
  SP&S: 116
  UP: 118
  WP: 123
F7
  ATSF: 31, 32, 34

ACL: 36
B&O: 38, 39
B&LE: 42
CP: 46
C&O: 20
C&NW: 50, 52, 53
CB&Q: 20
CGW: 56
CMStP&P: 60
CRI&P: 63
Clinchfield: 64
GN: 76
L&N: 84, 85, 86
NYC: 95
Reading: 104
SLSF: 106
SP: 113, 114
T&P: 117
Wabash: 119, 120
WM: 121

F9
   EMD: 70
   Erie Mining Co.: 72, 125
   NdeM: 23

FP7
   ACL: 37
   CP: 46, 47, 48
   CMStP&P: 59
   CRI&P: 61, 62
   Georgia RR: 74
   Reading: 25
   Soo: 107, 108

FP9
   CN: 26, 44

FL9
   NYNH&H: 27, 96, 97

Photographers

Abbey, Wallace W.: 32, 62
Akin, William A.: 85
Bain, Warren L. Jr.: 121
Baltimore & Ohio: 38
Bennett, James D.: 99
Benson, Ted: 28
Bessemer & Lake Erie: 42
Borcherding, Bob: 56
Botsko, S.: 65, 104
Bradley, R. Collins: 31
Brown, A. E.: 81
Canadian National: 44, 75
Canadian Pacific: 46
Chicago & Eastern Illinois: 49
Chicago & North Western: 53
Chicago, Burlington & Quincy: 19, 20, 54
Chicago Great Western: 57
Clarke, R. Jr.: 39
Cooke, M. B.: 35
Corey, G. C.: 45
Cutler, B. F.: 64
Denver & Rio Grande Western: 67
Electro-Motive: 4, 8, 10, 11, 12, 13, 15, 16, 17, 23, 25, 27, 58, 70
Emerick, Jack: 20
Farber, William L.: 83
Foster, R. J.: 106
Great Northern Railway: 76, 77
Harvey, H. P.: 43
Husa, William J. Jr.: 74
Illman, John C.: 47, 48
Kerrigan, Charles H.: 63
Kindig, R. H.: 68
Lamb, J. Parker: 84, 86, 120
LaVake, James G.: 109
Law, E. E. Jr.: 69
Marre, Louis A. (collection): 26, 91, 93, 94, 117
Marre, Louis A., and Mott, Gordon B., (collection): 17, 40
Matthews, Fred H. Jr.: 55
McClain, H. C.: 66
McCord, Henry J.: 59, 82
McLeod, Fred: 103
Meyer, Bruce R.: 119
Middleton, William D.: 36, 37, 108
Milner, Robert: 60, 79, 87, 88, 107
Missouri Pacific: 18
Morgan, David P. Library collection: 50, 72, 92
New York Central: 95
New York, New Haven & Hartford: 97
Nixon, R. V.: 6
Payne, R. H.: 40
Plummer, R. S.: 89, 106
Rail Photo Service: 99, 123
Russell, Larry: 26
St. Louis-San Francisco Railway: 105
Santa Fe Railway: 6, 29, 34
Sharp, Allen A.: 41
Shaughnessy, Jim: 41, 71, cover
Sims, Donald: 61, 90, 101, 111, 116, 118
Sisk, George W.: 93
Southern Railway: 109, 110
Small, Charles S.: 98
Steinheimer, Richard: 30, 33, 100, 102, 113, 114
Stevenson, Wade J.: 78
Stringham, Paul: 79
Treloar, E.: 52
Western Pacific: 122
Wood, Don: 96, 103
Wolfe, Harry M.: 73
Zygmund, George: 92